PRACTICE WORKBOOK

TEACHER'S EDITION

On My Own

Harcourt Brace & Company
Orlando • Atlanta • Austin • Boston • San Francisco • Chicago • Dallas • New York • Toronto • London
http://www.hbschool.com

Printed in the United States of America

ISBN 0-15-307942-8

1 2 3 4 5 6 7 8 9 10 082 2000 99 98 97

CONTENTS

CHAPTER 23 Measuring Solids

CHAPTER 24 Algebra: Number Relationships

CHAPTER 25 Operations with Integers

CHAPTER 26 Algebra: Equations and Relations

CHAPTER 27 Geometric Patterns

CHAPTER 28 Patterns and Operations

Name ____________________

LESSON 1.1

Whole Numbers and Decimals

Write the value of the digit 6.

1. 26.8 — 6 ones
2. 600 — 6 hundreds
3. 78.056 — 6 thousandths
4. 560,003 — 6 ten thousands

Write in words the value of the underlined digit.

5. 5<u>7</u>1 — 7 tens; 70
6. 9<u>2</u>,114.6 — 2 thousands; 2,000
7. 458.03<u>5</u> — 5 thousandths; 0.005
8. 8<u>6</u>2,077 — 6 ten thou- sands; 60,000

Write the number in words.

9. 12.7 — twelve and seven tenths
10. 254.91 — two hundred fifty-four and ninety-one hundredths
11. 4.814 — four and eight hundred fourteen thousandths

Write the number in standard form.

12. three thousand, forty-nine — 3,049
13. sixty thousand, three hundred five — 60,305
14. twenty-eight hundredths — 0.28
15. seventy-five and six tenths — 75.6

Mixed Applications

16. Over a holiday weekend, a total of six thousand, four hundred eighty-seven cars passed through the midtown tunnel. Write the number in standard form.

    6,487

17. The zigot factory made nine million, six hundred twenty-two thousand, seventeen zigots during the month of June. Write the number in standard form.

    9,622,017

18. A rectangle has a length of 6 in. and a width of 4 in. Find the area of the rectangle in square inches.

    24 in.^{2}

19. A scientist discovered that a plant grew 0.0027 cm over a 24-hr period. Use words to write the plant's growth.

    twenty-seven ten-thousandths

Use with text pages 16–17.

Name ______________________________

Comparing and Ordering

Compare the numbers. Write <, >, or =.

1. 15.4 __>__ 14.5
2. 5.67 __<__ 5.76
3. 43.90 __=__ 43.9
4. 7.91 __<__ 9.17
5. 765.28 __>__ 762.58
6. 0.234 __<__ 2.304

Write the numbers in order from least to greatest. Use <.

7. 3,224; 2,432; 3,422

2,432 < 3,224 < 3,422

8. 88.5; 85.8; 58.8

58.8 < 85.8 < 88.5

9. 6.21; 6.02; 6.12

6.02 < 6.12 < 6.21

Write the numbers in order from greatest to least. Use >.

10. 0.005; 0.500; 0.050

0.500 > 0.050 > 0.005

11. 317.8; 318.7; 371.8

371.8 > 318.7 > 317.8

12. 16.04; 14.6; 16.4

16.4 > 16.04 > 14.6

Mixed Applications

For Problems 13–16, use the table. It shows the average points scored by the top eight NBA scorers for one basketball season.

TOP NBA SCORERS			
Player	**Average**	**Player**	**Average**
Charles Barkley	25.6	Shaquille O'Neal	23.4
Joe Dumars	23.5	Hakeem Olajuwon	26.1
Patrick Ewing	24.2	Karl Malone	27.0
Michael Jordan	32.6	Dominique Wilkins	29.9

13. Compare the averages of Shaquille O'Neal, Patrick Ewing, Joe Dumars, and Charles Barkley. Write the numbers in order from least to greatest. Use <.

23.4 < 23.5 < 24.2 < 25.6

14. Compare the averages of Karl Malone, Michael Jordan, Hakeem Olajuwon, and Dominique Wilkins. Write the numbers in order from greatest to least. Use >.

32.6 > 29.9 > 27.0 > 26.1

15. What is the value of the digit 2 in the greatest average listed in the table?

2 ones; 2

16. Use words to write Shaquille O'Neal's average for the season.

twenty-three and four tenths

Use with text pages 18–19.

Name ______________________________

Decimals and Fractions

Use place value to change the decimal to a fraction.

1. 0.5 $\frac{5}{10}$
2. 0.14 $\frac{14}{100}$
3. 0.06 $\frac{6}{100}$
4. 0.83 $\frac{83}{100}$
5. 0.29 $\frac{29}{100}$
6. 0.62 $\frac{62}{100}$
7. 0.317 $\frac{317}{1,000}$
8. 0.8053 $\frac{8,053}{10,000}$
9. 0.955 $\frac{955}{1,000}$
10. 0.125 $\frac{125}{1,000}$

Write as a decimal.

11. $\frac{7}{10}$ **0.7**
12. $\frac{54}{100}$ **0.54**
13. $\frac{1}{8}$ **0.125**
14. $\frac{4}{5}$ **0.8**
15. $\frac{3}{16}$ **0.1875**
16. $\frac{91}{1,000}$ **0.091**
17. $\frac{3}{4}$ **0.75**
18. $\frac{1}{5}$ **0.2**
19. $\frac{19}{10,000}$ **0.0019**
20. $\frac{5}{16}$ **0.3125**
21. $\frac{32}{1,000}$ **0.032**
22. $\frac{5}{8}$ **0.625**
23. $\frac{156}{10,000}$ **0.0156**
24. $\frac{9}{16}$ **0.5625**
25. $\frac{89}{10,000}$ **0.0089**

Write < or >.

26. $\frac{3}{10} < \frac{2}{5}$
27. $\frac{11}{20} < \frac{7}{8}$
28. $\frac{91}{100} > \frac{11}{16}$
29. $0.24 < \frac{3}{4}$
30. $0.18 > \frac{7}{50}$
31. $0.04 < \frac{4}{10}$
32. $0.19 < \frac{1}{5}$
33. $0.459 > \frac{7}{20}$
34. $0.1 < \frac{15}{16}$

Mixed Applications

35. Rhea needs to know the decimal equivalent of $\frac{3}{8}$ to find the cost of three sticks of gum. Change the fraction to a decimal.

0.375

36. Tia needs to know the decimal equivalent of $\frac{5}{8}$ to find the cost of five slices of pizza. Change the fraction to a decimal.

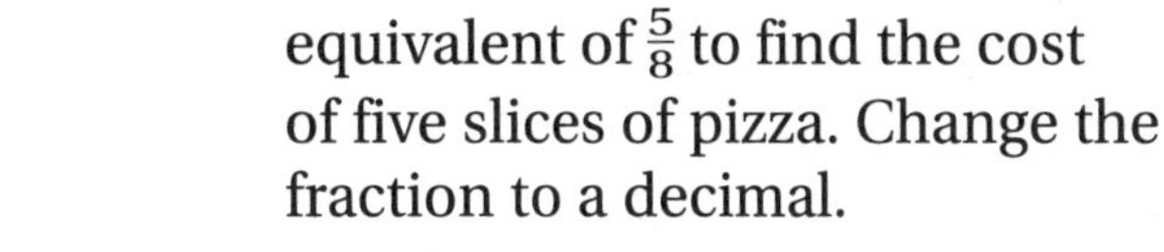

0.625

37. The batting averages of three players are 0.172, 0.223, and 0.194. Write the averages in order from greatest to least. Use >.

$0.223 > 0.194 > 0.172$

38. Todd and Jim drove home for the holidays. Todd drove $\frac{3}{8}$ of the distance. Jim drove $\frac{10}{16}$ of the distance. Who drove more miles? Explain.

Jim; $\frac{3}{8} < \frac{10}{16}$

Name ______________________

LESSON 1.4

Exponents

Vocabulary

Complete using *exponent* or *base*.

1. A(n) ___exponent___ shows how many times a number called the ___base___ is used as a factor.

Write in exponent form.

2. $5 \times 5 \times 5 \times 5$ ___5^4___

3. $10 \times 10 \times 10 \times 10 \times 10$ ___10^5___

4. 18×18 ___18^2___

Find the value.

5. 8^2 ___64___

6. 10^6 ___1,000,000___

7. 4^3 ___64___

8. 1^{18} ___1___

9. 2^6 ___64___

10. 6^4 ___1,296___

11. 11^2 ___121___

12. 10^3 ___1,000___

13. 15^1 ___15___

14. 30^2 ___900___

Express with an exponent and the given base.

15. 125, base 5 ___5^3___

16. 10,000, base 10 ___10^4___

17. 256, base 4 ___4^4___

18. 729, base 9 ___9^3___

Mixed Applications

19. Rod has a job that pays $5.00 on the first day of work. Then, for each day after the first, he receives double the preceding day's wage. Using exponent form, write the number of dollars he will receive on the eighth day.

20. Max has 6 cartons. In each carton, he places 6 bags. In each of the bags, he places 6 cookies. How many cookies are contained in the cartons? Use an exponent to write the answer.

___6^3___

21. While exercising, Maria jogged $\frac{3}{8}$ mi and walked $\frac{2}{5}$ mi. Did she jog or walk the greater distance? Explain.

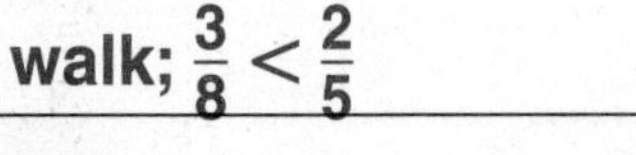

22. Three pieces of wire measure 28.3 cm, 23.8 cm, and 28.5 cm. Write the lengths in order from least to greatest. Use $<$.

___$23.8 \text{ cm} < 28.3 \text{ cm} < 28.5 \text{ cm}$___

Use with text pages 24–25.

Name ______________________________

Integers

Vocabulary

Write the correct letter from Column 2.

Column 1

c 1. integers

d 2. negative integers

a 3. opposites

b 4. positive integers

Column 2

a. are an equal distance from zero on the number line

b. are greater than zero

c. are either positive, negative, or zero

d. are less than zero

Compare the integers. Use $<$ or $>$.

5. $^{-}6$ $<$ 0 **6.** $^{+}11$ $>$ $^{-}12$ **7.** $^{-}2$ $>$ $^{-}8$ **8.** $^{+}17$ $>$ $^{-}21$ **9.** 0 $>$ $^{-}9$

Order the integers from least to greatest. Use $<$.

10. $^{-}2, ^{-}8, 0, ^{+}1$

$^{-}8 < ^{-}2 < 0 < ^{+}1$

11. $^{-}9, ^{+}11, ^{+}6, ^{-}1$

$^{-}9 < ^{-}1 < ^{+}6 < ^{+}11$

12. $^{+}5, ^{-}5, ^{+}7, ^{-}7$

$^{-}7 < ^{-}5 < ^{+}5 < ^{+}7$

13. $^{-}4, 0, ^{-}6, ^{+}1$

$^{-}6 < ^{-}4 < 0 < ^{+}1$

14. $^{-}11, ^{-}1, ^{+}1, ^{+}11$

$^{-}11 < ^{-}1 < ^{+}1 < ^{+}11$

15. $^{-}9, 0, ^{+}5, ^{-}3$

$^{-}9 < ^{-}3 < 0 < ^{+}5$

Name the opposite of the given integer.

16. $^{-}5$ — $^{+}5$

17. $^{+}13$ — $^{-}13$

18. $^{+}21$ — $^{-}21$

19. $^{-}19$ — $^{+}19$

20. $^{-}25$ — $^{+}25$

Mixed Applications

21. A starfish is 12 ft below sea level. A stingray is 8 ft below sea level. Which sea creature is closer to sea level?

stingray

22. The temperature in Jackson is $^{-}4$°F. The temperature in Paoli is $^{-}9$°F. Which city has the lower temperature?

Paoli

23. Jon has 5 containers. He put 5 envelopes in each container. Each envelope holds 5 baseball cards. How many cards are in the containers? Use an exponent to write the answer.

24. The test averages of three students are 89.7, 97.8, and 79.8. Write the averages in order from greatest to least. Use $>$.

$97.8 > 89.7 > 79.8$

Name ____________________

LESSON 2.1

Mental Math for Addition and Subtraction

Vocabulary

Match each term in Column A with its definition in Column B.

Column A

b 1. Associative Property

c 2. Commutative Property

a 3. compensation

Column B

a. Changing one addend and adjusting the other addend to keep the balance.

b. Addends can be grouped differently and produce the same sum.

c. Numbers can be added in any order without changing the sum.

Use mental math to add.

4. 12 + 7 + 18 37
5. 20 + 6 + 14 40
6. 9 + 4 + 11 24
7. 2 + 31 + 9 + 8 50
8. 23 + 18 + 17 + 12 70
9. 39 + 6 + 14 + 11 70

Use compensation to add.

10. 37 + 14 51
11. 59 + 26 85
12. 49 + 14 63
13. 56 + 15 71
14. 32 + 35 67
15. 41 + 18 59

Use compensation to subtract.

16. 65 − 23 42
17. 78 − 45 33
18. 37 − 11 26
19. 96 − 25 71
20. 88 − 54 34
21. 74 − 23 51

Mixed Applications

Solve.

For Problem 22, use the table at the right.

22. Which two grades collected a total of 60 cans?

grades 3 and 6

23. The Stambaughs have a rectangular pool that measures 20 ft by 40 ft. What is the size of a pool cover that will cover the pool exactly?

800 ft^2

CANS COLLECTED			
Grade 3	Grade 4	Grade 5	Grade 6
18	24	26	42

24. Brad needs 100 signatures on a petition. On Monday he got 23 signatures, on Tuesday he got 12, and on Wednesday and Thursday he got 15 each. How many more signatures does he need?

35 more signatures

Use with text pages 34–35.

Name ______________________________

Problem-Solving Strategy

Using Guess and Check to Add and Subtract

Guess and check to solve.

1. Ryan bought a total of 40 juice boxes. He bought 8 more boxes of apple juice than of grape juice. How many of each kind did he buy?

   24 apple juice, 16 grape juice

2. The perimeter of a rectangular garden is 56 ft. the length is 4 ft more than the width. What are the dimensions of the garden?

   l = 16 ft; w = 12 ft

3. The Hawks soccer team played a total of 24 games. They won 6 more games than they lost, and they tied 2 games. How many games did they win?

   14 games

4. Rico collected a total of 47 rocks. He gathered 5 more jagged rocks than smooth rocks. How many jagged and smooth rocks did he collect?

   26 jagged rocks, 21 smooth rocks

Mixed Applications

Solve.

CHOOSE A STRATEGY

- Guess and Check
- Solve a Simpler Problem
- Write an Equation
- Find a Pattern
- Act It Out

5. Matt has earned $75. To buy a bicycle, he needs twice that amount plus $30. How much does the bicycle cost?

   $180

6. The perimeter of a rectangular lot is 190 ft. The width of the lot is 15 ft more than the length. What are the dimensions of the lot?

   w = 55 ft; l = 40 ft

7. The Chen family leaves for vacation at 10:00 A.M. They drive at an average speed of 52 mi per hour. They stop for lunch at 1:00 P.M. How many miles have they traveled?

   156 mi

8. Valley High School's football team played a total of 16 games. They won twice as many games as they lost. If they tied one game, how many games did the team win?

   10 games

9. Marc spent 30 min on math homework and 45 min on language homework. Then he played computer games for 20 min. He started at 6:30 P.M. What time did he finish?

   8:05 P.M.

10. Rhea spent a total of $43.00. She bought a sweater for $19.75, a scarf for $4.50, a blouse for $12.75, and some socks for $2.00 a pair. How many pairs of socks did she buy?

    3 pairs of socks

Name ______________________

LESSON 2.3

Mental Math Strategies for Multiplication

Vocabulary

Match each property in Column A with its definition in Column B.

Column A

c 1. Associative Property

e 2. Commutative Property

b 3. Distributive Property

d 4. Identity Property of One

a 5. Zero Property

Column B

a. The product of any factor and zero is zero.

b. Multiplying a factor's addends by a number, then adding the products, is the same as multiplying the factor by the number.

c. Factors can be grouped in any way without changing the product.

d. The product of any factor and 1 is the factor.

e. Factors can be multiplied in any order without changing the product.

Find the missing factor.

6. $56 \times 7 = (50 \times 7) + (\underline{6} \times 7)$

7. $15 \times \underline{48} = (15 \times 40) + (15 \times 8)$

8. $8 \times 93 = (8 \times \underline{90}) + (8 \times 3)$

9. $63 \times 21 = (\underline{63} \times 20) + (63 \times 1)$

Use the Distributive Property and mental math to find the product.

10. 18×6 108
11. 42×5 210
12. 17×4 68
13. 31×15 465
14. 86×3 258
15. 18×22 396
16. 47×3 141
17. 15×51 765

Use mental math to find the product.

18. $8 \times 3 \times 4$ 96
19. $2 \times 7 \times 5$ 70
20. $6 \times 9 \times 5$ 270
21. $10 \times 4 \times 7$ 280
22. 17×9 153
23. 61×6 366
24. 19×11 209
25. $4 \times 16 \times 5$ 320

Mixed Applications

26. The Sports Shack has 8 boxes of baseballs on each of 5 shelves. Each box holds one dozen balls. How many baseballs are there in all?

480 baseballs

27. Bill has earned $87. To buy a new stereo, he needs twice that amount plus $50. How much does the stereo cost?

$224

Use with text pages 38–39.

Name ____________________

Multiplication and Division

Find the product.

1. 46 × 12 = **552**
2. 230 × 15 = **3,450**
3. 417 × 40 = **16,680**
4. 2,515 × 52 = **130,780**
5. 387 × 66 = **25,542**
6. 217 × 154 = **33,418**
7. 6,903 × 627 = **4,328,181**
8. 582 × 316 = **183,912**
9. 6,148 × 744 = **4,574,112**
10. 8,132 × 915 = **7,440,780**

Find the quotient.

11. 4)96 **24**
12. 9)423 **47**
13. 19)361 **19**
14. 7)756 **108**
15. 32)450 **14 r2**
16. 12)1,740 **145**
17. 19)912 **48**
18. 22)5,412 **246**
19. 31)4,836 **156**
20. 17)5,865 **345**
21. 13)3,302 **254**
22. 37)1,075 **29 r2**
23. 45)17,145 **381**
24. 91)1,001 **11**
25. 23)4,720 **205 r5**

Find the quotient. Write the remainder as a fraction.

26. 6)45 **$7\frac{1}{2}$**
27. 14)550 **$39\frac{2}{7}$**
28. 18)459 **$25\frac{1}{2}$**
29. 41)13,210 **$322\frac{8}{41}$**
30. 55)33,125 **$602\frac{3}{11}$**

Mixed Applications

31. Tonya earns $43,800 a year as a city planner. How much does she earn each month?

$3,650

32. Jon spent $45 on a new pair of golf shoes, $5 for golf balls, and $16 on a hat. How much did he spend in all?

$66

33. Kyle is buying a used car. He has found two cars that he likes. The red car costs $3,709. A similar model in blue costs $3,079. Kyle decides to buy the less expensive car. Which car does he buy?

the blue car

34. Dave bought a bedroom set for $2,250. If he pays for it with 24 equal monthly payments, how much will each payment be?

$93.75

Name ____________________

LESSON 2.5

Using Estimation

Vocabulary

1. Numbers that divide without a remainder, are close to the actual numbers, and are easy to compute mentally are called **compatible numbers**.

2. When all addends are about the same, you can use **clustering** to estimate their sum.

Estimate the sum or difference. **Possible estimates are given.**

3. 2,489 + 1,601 + 2,109 = **6,000**

4. 398 + 415 + 368 = **1,200**

5. 4,723 + 2,198 = **6,800**

6. 7,132 + 6,594 + 7,301 = **21,000**

7. 5,401 + 9,188 = **14,500**

8. 478 − 26 = **450**

9. 263 − 211 = **50**

10. 5,877 − 5,318 = **600**

11. 8,528 − 6,491 = **2,000**

12. 8,903 − 4,575 = **4,300**

Estimate the product. **Possible estimates are given.**

13. 53 × 8 = **400**

14. 76 × 9 = **720**

15. 72 × 28 = **2,100**

16. 47 × 53 = **2,500**

17. 660 × 42 = **28,000**

18. 371 × 78 = **32,000**

19. 68 × 37 = **2,800**

20. 480 × 192 = **100,000**

21. 375 × 591 = **240,000**

22. 824 × 693 = **560,000**

Estimate the quotient. **Possible estimates are given.**

23. 331 ÷ 5 **70**

24. 643 ÷ 9 **70**

25. 1,827 ÷ 59 **30**

26. 5,543 ÷ 77 **70**

27. 9,165 ÷ 28 **300**

28. 6,281 ÷ 875 **7**

29. 7,118 ÷ 614 **12**

30. 8,215 ÷ 897 **9**

Mixed Applications

31. There are 187 seats in a local movie theater. The theater has been sold out for the past 18 shows. About how many people attended the shows?

about 4,000 people

32. Rhonda earned $371.09 this week. Her brother Rod earned $397.01. Who earned more money? How much more?

Rod; $25.92 more

Use with text pages 46–49.

Name ______________________

LESSON 3.1

Adding Decimals

Estimate the sum. **Possible estimates are given.**

1. 1.6 + 3.3 — **5**
2. 0.2 + 0.9 — **1**
3. 8.1 + 7.5 — **16**
4. 5.16 + 0.58 — **6**
5. 9.14 + 8.68 — **18**
6. 7.58 + 3.25 — **11**
7. 0.291 + 1.833 — **2**
8. 4.693 + 9.711 — **15**

Find the sum.

9. 0.34 + 8.19 — **8.53**
10. 6.92 + 3.55 — **10.47**
11. 0.418 + 3.291 — **3.709**
12. 8.93 + 2.60 — **11.53**
13. 4.89 + 2.45 — **7.34**
14. 0.68 + 7.12 — **7.8**
15. 1.681 + 2.899 — **4.58**
16. 7.86 + 4.20 — **12.06**
17. 1.246 + 2.081 — **3.327**
18. 59.328 + 1.294 — **60.622**
19. 5.0804 + 23.7381 — **28.8185**
20. 446.09 + 811.36 + 73.52 — **1,330.97**
21. 8.71 + 13.99 + 67.2 — **89.9**
22. 23.75 + 873.33 + 2,586.02 — **3,483.10**
23. 3.056 + 28.1174 + 1,691.396 + 44.21 — **1,766.7794**

Mixed Applications

24. The Garden Club has raised $213.42 for a new garden in the park. Plants for the garden cost $178.95, fencing costs $87.99, and fertilizer costs $44.15. About how much more money does the club need to cover the costs?

    about $100

25. Ryan read 8 pages of a book today. He plans to increase the number of pages he reads each day by 4. How many pages will he read four days from now?

    24 pages

26. Beth needs soccer equipment. Shoes cost $39.95, shorts cost $19.95, and socks cost $2.79. How much money will she need to buy the equipment?

    $62.69

27. This year, 432 students signed up to play intramural sports. Each team needs 18 players. How many teams are there?

    24 teams

Name ______________________________

Subtracting Decimals

Find the difference.

1. 8.7 − 4.2 **4.5**
2. 13.2 − 5.9 **7.3**
3. 5.41 − 1.36 **4.05**
4. $15.93 − $7.08 **$8.85**
5. 5.962 − 1.748 **4.214**
6. 4.036 − 2.751 **1.285**
7. 8.1163 − 3.0948 **5.0215**
8. 17.1053 − 12.7559 **4.3494**
9. 15.082 − 4.19 **10.892**
10. 18.1429 − 6.204 **11.9389**
11. 14.16 − 6.385 **7.775**
12. 45.324 − 7.0871 **38.2369**
13. 16.076 − 3.28 **12.796**
14. 5 − 3.218 = **1.782**
15. 54.08 − 29.7561 = **24.3239**
16. 14 − 5.39 = **8.61**
17. 631.5039 − 420.1417 = **211.3622**
18. 21 − 9.45 = **11.55**
19. 89.01 − 67.56 = **21.45**
20. 25.8 − 17.226 = **8.574**
21. 71.043 − 58.6492 = **12.3938**

Find the missing number.

22. 1.83 + **0.98** = 2.81
23. 10.13 + **4.96** = 15.09
24. 12.379 + **11.021** = 23.4
25. 16.007 + **20.576** = 36.583

Mixed Applications

26. Mel uses 2.8 lb from a 10-lb bag of potatoes to make hash browns. How many pounds of potatoes are left?

7.2 lb

27. The distance from Joan's house to school is 6.3 mi. The distance from Rick's house to school is 1.9 mi. How much closer to the school is Rick's house?

4.4 mi closer

28. Zack has $6.00 for grocery shopping. Juice costs $2.29, bread costs $1.29, and eggs cost $1.09. How much more money does he need to get 2 of each item?

$3.34 more

29. Carla does 22 sit-ups a day. She plans to increase her exercise by 4 sit-ups a day. How many sit-ups will she be doing at the end of 3 days?

34 sit-ups

Use with text pages 56–57.

Name ______________________

Multiplying Decimals

Estimate the product. **Possible estimates are given.**

1. 6.3 × 0.75 — **6**
2. 9.7 × 48.8 — **490**
3. 5.96 × 62.15 — **360**
4. 37.6 × 8.3 — **320**
5. 32.08 × 7.3 — **210**
6. 428.9 × 5.6 — **2,400**
7. 897.35 × 5.3 — **4,500**
8. 186.45 × 9.6 — **2,000**

Place the decimal point in the product.

9. 6.17 × 8.2 = 50594 — **50.594**
10. 24.01 × 8.51 = 2043251 — **204.3251**
11. 8.94 × 5.27 = 471138 — **47.1138**
12. 8.04 × 1.7 = 13668 — **13.668**
13. 19.6 × 5.8 = 11368 — **113.68**
14. 30.7 × 8.33 = 255731 — **255.731**

Find the product.

15. 5 × 0.9 — **4.5**
16. 9 × 1.2 — **10.8**
17. 4 × 3.47 — **13.88**
18. $18.93 × 7 — **$132.51**
19. $5.55 × 9 — **$49.95**
20. 5 × 2.89 — **14.45**
21. $31.82 × 4 — **$127.28**
22. 4.61 × 8 — **36.88**
23. $2.49 × 6 — **$14.94**
24. 35.98 × 6.3 — **226.674**
25. 73.02 × 9.1 — **664.482**
26. 8.5 × 16.03 — **136.255**

Mixed Applications

27. The Flower Club is selling plants for $7.95 each. If the club sells a total of 285 plants, how much money will they make?

    $2,265.75

28. Ronnie is selling handmade baby sweaters. She charges $29.50 for one sweater. If she sells 32 sweaters, how much money will she make?

    $944.00

29. This year 592 students signed up to play basketball. Each team needs 8 players. How many teams are there?

    74 teams

30. There are 104 students in the school band. They are traveling in cars to a performance. Each car will transport 4 band members. How many cars are needed?

    26 cars

Name ____________________

LESSON 3.4

Dividing Decimals

Rewrite the problem so that the divisor is a whole number.

1. $8.5 \div 2.3$ — **$85 \div 23$**
2. $6.4 \div 1.3$ — **$64 \div 13$**
3. $9.1 \div 0.15$ — **$910 \div 15$**
4. $33.17 \div 6.8$ — **$331.7 \div 68$**

Complete.

5. $56.8 \div 0.8 = 568 \div$ **8**
6. $7.21 \div 0.03 =$ **721** $\div 3$
7. $4.12 \div 2.3 = 41.2 \div$ **23**

Find the quotient.

8. $36.9 \div 3$ — **12.3**
9. $22.4 \div 7$ — **3.2**
10. $37.5 \div 5$ — **7.5**
11. $89.6 \div 8$ — **11.2**

12. $14\overline{)78.4}$ **5.6**
13. $40\overline{)6.8}$ **0.17**
14. $13\overline{)150.8}$ **11.6**
15. $70\overline{)23.8}$ **0.34**

16. $5.32 \div 0.7$ — **7.6**
17. $1.88 \div 0.4$ — **4.7**
18. $2.12 \div 0.2$ — **10.6**
19. $5.4 \div 0.08$ — **67.5**

20. $7.54\overline{)24.882}$ **3.3**
21. $12.6\overline{)806.4}$ **64**
22. $0.91\overline{)6.734}$ **7.4**
23. $10.9\overline{)81.75}$ **7.5**

24. $2.9\overline{)0.3335}$ **0.115**
25. $0.18\overline{)64.296}$ **357.2**
26. $12.3\overline{)84.87}$ **6.9**
27. $8.7\overline{)53.244}$ **6.12**

Mixed Applications

28. Ryan buys 6 lb of mixed nuts for \$16.74. How much does 1 lb of nuts cost?

\$2.79

29. Jon buys golf balls for \$12.79. He gives the clerk a \$20 bill. How much change does Jon get?

\$7.21

30. James is saving \$15.75 every week to buy a plane ticket that costs \$141.75. In how many weeks will he have enough money saved?

9 weeks

31. Michele jogged $1\frac{1}{2}$ mi this week. She plans to increase the distance by $\frac{1}{2}$ mi every week. How far will she jog 5 weeks from now?

4 mi

Use with text pages 64–67.

Name ____________________

LESSON 4.1

Multiples and Factors

Vocabulary

Complete.

1. Numbers that have more than two factors are called composite numbers.

2. Numbers with only two factors are called prime numbers.

Name the first four multiples.

3. 12 — 12, 24, 36, 48
4. 20 — 20, 40, 60, 80
5. 18 — 18, 36, 54, 72
6. 13 — 13, 26, 39, 52

Find the missing multiples.

7. 6, 12, 18, 24, 30
8. 14, 28, 42, 56, 70
9. 13, 26, 39, 52, 65

Write the factors.

10. 28 — 1, 2, 4, 7, 14, 28
11. 51 — 1, 3, 17, 51
12. 37 — 1, 37
13. 72 — 1, 2, 3, 4, 6, 8, 9, 12, 18, 24, 36, 72

Write *P* for *prime* or *C* for *composite.*

14. 54 C
15. 53 P
16. 42 C

Mixed Applications

17. Fran has 36 calendars. Each person in her office can have 4 calendars. How many people work in Fran's office? Is that number a factor of 36?

9; yes

18. Shawna is making a poster that measures 14 in. by 20 in. What is the area of the poster?

280 $in.^2$

19. Pete played tennis every sixth day in July, beginning on July 6. How many times did he play in July? on what dates?

5 times; July 6, July 12, July 18, July 24, July 30

20. Grace wants to make equal-size packages of pens from 32 pens. In how many ways can she do that? How many pens would be in each package?

5 ways: packages of 1, 2, 4, 8, or 16

Name ____________________

LESSON 4.2

Prime Factorization

Vocabulary

1. Write *true* or *false*. Prime factorization is writing a composite number as the product of composite factors. **false**

Use division to find the prime factors. Write the prime factorization.

2. 28 — $2 \times 2 \times 7$
3. 50 — $2 \times 5 \times 5$
4. 76 — $2 \times 2 \times 19$
5. 108 — $2 \times 2 \times 3 \times 3 \times 3$

Use a factor tree to find the prime factors. Write the prime factorization in exponent form. **Check students' factor trees.**

6. 55 — 5×11
7. 120 — $2 \times 2 \times 2 \times 3 \times 5$; $2^3 \times 3 \times 5$
8. 92 — $2 \times 2 \times 23$; $2^2 \times 23$

Solve for *n* to complete the prime factorization.

9. $n \times 17 = 51$ **3**
10. $3^n \times 2 = 18$ **2**
11. $2 \times 2 \times 2 \times n = 40$ **5**

Mixed Applications

12. Beth has 6 coins in her pocket. The total amount is $1.00. She has no pennies. What are the coins?

3 quarters, 2 dimes, 1 nickel or 5 dimes and 1 half-dollar or 1 half-dollar, 1 quarter, 1 dime, 3 nickels

13. There are 24 students in Mrs. Garcia's class. She wants to divide the class evenly into groups of at least 4 students. Write the ways in which she can divide the class.

2 groups of 12, 4 groups of 6, 6 groups of 4, or 3 groups of 8

14. The two prime factors of a number are greater than 12 but less than 25. The number is greater than 300 but less than 400. What is the number? What are the factors?

323; 17 and 19, or 391; 17 and 23

15. The three prime factors of a number are all less than 10. The number is greater than 100. If no factor is repeated, what is the number? What are the factors?

105; 3, 5, and 7

Name ____________________

LCM and GCF

Vocabulary

Complete.

1. The smallest of the common multiples is called the

   least common multiple, or LCM.

2. The largest of the common factors is called the

   greatest common factor, or GCF.

Find the LCM for each set of numbers.

3. 12, 18 — 36
4. 7, 14 — 14
5. 16, 20 — 80
6. 4, 5, 6 — 60
7. 2, 6, 7 — 42

Find the GCF for each set of numbers.

8. 15, 45 — 15
9. 6, 14 — 2
10. 24, 40 — 8
11. 8, 12, 52 — 4
12. 16, 24, 32 — 8

Find the common prime factors. Then find the GCF.

13. 15, 50 — 5; 5
14. 16, 24 — $2 \times 2 \times 2$; 8
15. 49, 70 — 7; 7
16. 45, 108 — 3×3; 9
17. 18, 36 — $2 \times 3 \times 3$; 18

Mixed Applications

18. Walt has 51 football stickers and 68 baseball stickers. He will put them on cards that all have the same number of stickers. What is the greatest number of cards Walt can make?

    17 cards

19. Hot dogs come in packages of 8, and rolls come in packages of 6. What is the smallest number of each that you can buy so there are no extras?

    3 packages of hot dogs,
    4 packages of rolls

20. Maya has 65 dimes and 104 pennies. She will put them in packages that are all the same. What is the greatest number of packages she can make?

    13 packages

21. Rob jogged on the 4th, 8th, and 12th of the month. If he continues jogging in this pattern, what will be the next two days that Rob jogs?

    16th and 20th

Name ______________________

Fractions in Simplest Form

Vocabulary

Complete.

1. When the numerator and denominator of a fraction have no common factor other than 1, the fraction is in **simplest form**.

Write the common factors and the GCF of the numerator and denominator.

2. $\frac{8}{32}$ **1, 2, 4, 8; 8**
3. $\frac{10}{50}$ **1, 2, 5, 10; 10**
4. $\frac{2}{13}$ **1; 1**
5. $\frac{14}{49}$ **1, 7; 7**
6. $\frac{1}{19}$ **1; 1**
7. $\frac{12}{18}$ **1, 2, 3, 6; 6**
8. $\frac{25}{75}$ **1, 5, 25; 25**
9. $\frac{15}{40}$ **1, 5; 5**
10. $\frac{9}{54}$ **1, 3, 9; 9**
11. $\frac{6}{33}$ **1, 3; 3**

Write the fraction in simplest form.

12. $\frac{9}{36}$ **$\frac{1}{4}$**
13. $\frac{15}{50}$ **$\frac{3}{10}$**
14. $\frac{11}{121}$ **$\frac{1}{11}$**
15. $\frac{15}{36}$ **$\frac{5}{12}$**
16. $\frac{14}{28}$ **$\frac{1}{2}$**
17. $\frac{30}{66}$ **$\frac{5}{11}$**
18. $\frac{63}{72}$ **$\frac{7}{8}$**
19. $\frac{27}{81}$ **$\frac{1}{3}$**
20. $\frac{25}{65}$ **$\frac{5}{13}$**
21. $\frac{12}{42}$ **$\frac{2}{7}$**

Write the missing number.

22. $\frac{36}{72} = \frac{1}{\boxed{2}}$
23. $\frac{\boxed{50}}{75} = \frac{2}{3}$
24. $\frac{17}{\boxed{85}} = \frac{1}{5}$
25. $\frac{63}{84} = \frac{3}{\boxed{4}}$
26. $\frac{2}{\boxed{3}} = \frac{64}{96}$

Mixed Applications

27. Ryan has 6 Yankees baseball cards, 5 Dodgers cards, and 7 Braves cards. What fraction of the cards are Yankees cards? Write the fraction in simplest form.

    $\frac{1}{3}$

28. Wanda has 5 containers of orange juice, 11 of apple juice, and 4 of grape juice. What fraction of the containers are *not* orange juice? Write the fraction in simplest form.

    $\frac{3}{4}$

29. Ron rents a car for 6 days at $22.95 per day and $0.15 per mile. He travels 420 miles. What is his total cost?

    $200.70

30. Betty rents a VCR for 1 week at $5.95 per day. She also rents 3 tapes, each for $4.95 per week. What is her total cost?

    $56.50

Use with text pages 86–87.

Name ______________________

LESSON 4.5

Mixed Numbers and Fractions

Vocabulary

Complete.

1. A whole number and a fraction is called a(n) ___mixed number___.

Write the fraction as a mixed number or a whole number.

2. $\frac{20}{5}$ ___4___
3. $\frac{19}{4}$ ___$4\frac{3}{4}$___
4. $\frac{22}{7}$ ___$3\frac{1}{7}$___
5. $\frac{39}{10}$ ___$3\frac{9}{10}$___
6. $\frac{19}{10}$ ___$1\frac{9}{10}$___
7. $\frac{75}{15}$ ___5___
8. $\frac{44}{13}$ ___$3\frac{5}{13}$___
9. $\frac{50}{7}$ ___$7\frac{1}{7}$___
10. $\frac{63}{21}$ ___3___
11. $\frac{41}{8}$ ___$5\frac{1}{8}$___

Write the mixed number as a fraction.

12. $6\frac{2}{7}$ ___$\frac{44}{7}$___
13. $4\frac{6}{11}$ ___$\frac{50}{11}$___
14. $9\frac{2}{3}$ ___$\frac{29}{3}$___
15. $11\frac{1}{5}$ ___$\frac{56}{5}$___
16. $2\frac{2}{3}$ ___$\frac{8}{3}$___

Tell which are equivalent numbers in each set.

17. $8\frac{1}{5}, \frac{81}{5}, 41\frac{4}{5}, \frac{41}{5}$ ___$8\frac{1}{5}$ and $\frac{41}{5}$___
18. $\frac{22}{3}, 8\frac{2}{3}, \frac{19}{3}, 6\frac{1}{3}$ ___$\frac{19}{3}$ and $6\frac{1}{3}$___

Write the missing number.

19. $\frac{\boxed{54}}{11} = 4\frac{10}{11}$
20. $\frac{\boxed{97}}{12} = 8\frac{1}{12}$
21. $7\frac{\boxed{5}}{9} = \frac{68}{9}$
22. $9\frac{1}{2} = \frac{\boxed{19}}{2}$

Mixed Applications

23. Jon jogs $\frac{7}{4}$ of a mile. Kathy jogs $1\frac{3}{4}$ of a mile. Do they jog the same distance? Explain.

Yes; $\frac{7}{4} = 1\frac{3}{4}$.

24. Vicky has 32 pens. Each person in her office can take 4 pens. How many people work in her office?

8 people

25. Terri is making a quilt. She needs $\frac{4}{3}$ yd of fabric. If she buys $1\frac{1}{2}$ yd, will she have enough to make the quilt? Explain.

Yes. The quilt requires only $1\frac{1}{3}$ yd of fabric.

26. Chris has a dozen pieces of fruit. She has 4 oranges, 5 apples, and some bananas. What fraction of the fruit is bananas? Write the fraction in simplest form.

$\frac{1}{4}$

Name ____________________

LESSON 5.1

Adding and Subtracting Like Fractions

Find the sum or difference. Write the answer in simplest form.

1. $\frac{4}{9} + \frac{5}{9}$ **1**
2. $\frac{3}{8} + \frac{1}{8}$ $\frac{1}{2}$
3. $\frac{2}{7} + \frac{3}{7}$ $\frac{5}{7}$
4. $\frac{1}{4} + \frac{1}{4}$ $\frac{1}{2}$
5. $\frac{3}{10} + \frac{1}{10}$ $\frac{2}{5}$
6. $\frac{8}{9} - \frac{2}{9}$ $\frac{2}{3}$
7. $\frac{4}{5} - \frac{2}{5}$ $\frac{2}{5}$
8. $\frac{7}{8} - \frac{3}{8}$ $\frac{1}{2}$
9. $\frac{5}{6} - \frac{1}{6}$ $\frac{2}{3}$
10. $\frac{8}{11} - \frac{4}{11}$ $\frac{4}{11}$
11. $\frac{6}{7} + \frac{3}{7}$ $1\frac{2}{7}$
12. $\frac{9}{10} + \frac{3}{10}$ $1\frac{1}{5}$
13. $\frac{1}{6} + \frac{5}{6}$ **1**
14. $\frac{3}{4} + \frac{1}{4}$ **1**
15. $\frac{7}{11} + \frac{5}{11}$ $1\frac{1}{11}$
16. $\frac{7}{10} - \frac{2}{10}$ $\frac{1}{2}$
17. $\frac{11}{14} - \frac{3}{14}$ $\frac{4}{7}$
18. $\frac{17}{20} - \frac{7}{20}$ $\frac{1}{2}$
19. $\frac{13}{15} - \frac{4}{15}$ $\frac{3}{5}$
20. $\frac{5}{8} - \frac{3}{8}$ $\frac{1}{4}$
21. $\frac{4}{17} + \frac{6}{17}$ $\frac{10}{17}$
22. $\frac{8}{9} - \frac{5}{9}$ $\frac{1}{3}$
23. $\frac{4}{11} + \frac{7}{11}$ **1**
24. $\frac{17}{18} - \frac{3}{18}$ $\frac{7}{9}$
25. $\frac{7}{12} + \frac{1}{12}$ $\frac{2}{3}$
26. $\frac{4}{16} + \frac{8}{16}$ $\frac{3}{4}$
27. $\frac{11}{15} - \frac{5}{15}$ $\frac{2}{5}$
28. $\frac{8}{9} + \frac{1}{9}$ **1**
29. $\frac{10}{12} - \frac{9}{12}$ $\frac{1}{12}$
30. $\frac{8}{15} + \frac{4}{15}$ $\frac{4}{5}$

Mixed Applications

31. Josh has read $\frac{2}{5}$ of a book. What part of the book must he still read?

    $\frac{3}{5}$ of the book

32. Sam eats $\frac{2}{9}$ of a cake. Ted eats $\frac{4}{9}$ of the same cake. How much of the cake do they eat? How much is left?

    $\frac{2}{3}$ eaten; $\frac{1}{3}$ left

33. A number has 3 prime factors. The number is greater than 25 but less than 35. If no factor is repeated, what is the number? What are the factors?

    30; $2 \times 3 \times 5$

34. Rod has 8 coins in his pocket. He has no half dollars. The value of the coins is $1.17. What are the coins?

    4 quarters, 1 dime, 1 nickel, 2 pennies

Use with text pages 94–95.

Name ______________________

Adding and Subtracting Unlike Fractions

Write the addition or subtraction problem and then solve.

1.

$\frac{1}{5}$		$\frac{1}{5}$		$\frac{1}{5}$		$\frac{1}{10}$
$\frac{1}{10}$	$\frac{1}{10}$	$\frac{1}{10}$	$\frac{1}{10}$	$\frac{1}{10}$	$\frac{1}{10}$	$\frac{1}{10}$

$\frac{3}{5} + \frac{1}{10} = \frac{7}{10}$

2.

$\frac{1}{8}$	$\frac{1}{8}$	$\frac{1}{8}$	$\frac{1}{8}$	$\frac{1}{8}$
$\frac{1}{4}$				

$\frac{5}{8} - \frac{1}{4} = \frac{3}{8}$

3.

$\frac{1}{3}$			$\frac{1}{3}$			$\frac{1}{9}$
$\frac{1}{9}$	$\frac{1}{9}$	$\frac{1}{9}$	$\frac{1}{9}$	$\frac{1}{9}$	$\frac{1}{9}$	$\frac{1}{9}$

$\frac{2}{3} + \frac{1}{9} = \frac{7}{9}$

Draw a diagram to find each sum or difference. **Check students' diagrams.**

4. $\frac{1}{2} - \frac{1}{6}$ **$\frac{1}{3}$**

5. $\frac{3}{10} + \frac{2}{5}$ **$\frac{7}{10}$**

6. $\frac{5}{6} - \frac{1}{4}$ **$\frac{7}{12}$**

7. $\frac{2}{9} + \frac{1}{3}$ **$\frac{5}{9}$**

8. $\frac{7}{8} - \frac{1}{4}$ **$\frac{5}{8}$**

9. $\frac{1}{3} + \frac{1}{2}$ **$\frac{5}{6}$**

10. $\frac{7}{10} - \frac{3}{5}$ **$\frac{1}{10}$**

11. $\frac{1}{3} + \frac{1}{6}$ **$\frac{1}{2}$**

12. $\frac{8}{9} - \frac{2}{3}$ **$\frac{2}{9}$**

Mixed Applications

13. Spencer's backyard measures 12 yd by 8 yd. What is the area of his backyard?

96 sq yd

14. Chris eats $\frac{1}{8}$ of a pizza. Ryan eats $\frac{3}{4}$ of the same pizza. How much of the pizza do they eat? How much is left?

$\frac{7}{8}$ eaten; $\frac{1}{8}$ left

15. Joan walked $\frac{1}{5}$ mi to her friend's house. She then walked $\frac{1}{4}$ mi to school. What is the total distance that Joan walked?

$\frac{9}{20}$ mi

16. In Mr. Sims's math class, $\frac{2}{5}$ of the students are wearing blue shirts and $\frac{3}{10}$ of the students are wearing white shirts. What part of the class is wearing *neither* blue nor white shirts?

$\frac{3}{10}$ of the class

Name ______________________

LESSON 5.3

Adding Unlike Fractions

Vocabulary

1. To add unlike fractions, you write equivalent fractions by using the

least common denominator.

Write equivalent fractions with the LCD.

2. $\frac{3}{8} + \frac{1}{2}$ **$\frac{3}{8} + \frac{4}{8}$**

3. $\frac{3}{4} + \frac{1}{6}$ **$\frac{9}{12} + \frac{2}{12}$**

4. $\frac{7}{8} + \frac{1}{4}$ **$\frac{7}{8} + \frac{2}{8}$**

5. $\frac{2}{3} + \frac{4}{5}$ **$\frac{10}{15} + \frac{12}{15}$**

6. $\frac{7}{10} + \frac{4}{5}$ **$\frac{7}{10} + \frac{8}{10}$**

7. $\frac{7}{12} + \frac{2}{3}$ **$\frac{7}{12} + \frac{8}{12}$**

8. $\frac{8}{9} + \frac{1}{3}$ **$\frac{8}{9} + \frac{3}{9}$**

9. $\frac{11}{15} + \frac{3}{5}$ **$\frac{11}{15} + \frac{9}{15}$**

Find the sum. Write the answer in simplest form.

10. $\frac{1}{3} + \frac{2}{9}$ **$\frac{5}{9}$**

11. $\frac{1}{7} + \frac{1}{2}$ **$\frac{9}{14}$**

12. $\frac{2}{3} + \frac{5}{12}$ **$\frac{13}{12}$, or $1\frac{1}{12}$**

13. $\frac{3}{10} + \frac{1}{2}$ **$\frac{4}{5}$**

14. $\frac{1}{3} + \frac{7}{9}$ **$\frac{10}{9}$, or $1\frac{1}{9}$**

15. $\frac{4}{5} + \frac{1}{6}$ **$\frac{29}{30}$**

16. $\frac{3}{8} + \frac{1}{3}$ **$\frac{17}{24}$**

17. $\frac{4}{9} + \frac{1}{2}$ **$\frac{17}{18}$**

18. $\frac{1}{10} + \frac{1}{4}$ **$\frac{7}{20}$**

19. $\frac{3}{12} + \frac{1}{4}$ **$\frac{1}{2}$**

Mixed Applications

20. Ron is mixing salad dressing. He combines $\frac{1}{2}$ c oil with $\frac{1}{4}$ c vinegar. What is the total amount of oil and vinegar he uses?

$\frac{3}{4}$ c

21. Marie is making a birdhouse. She needs $\frac{2}{3}$ ft of plywood for the roof and $\frac{5}{6}$ ft of plywood for the sides. How much plywood does she need for the birdhouse?

$1\frac{1}{2}$ ft

22. The Garden Club sold plants to raise money for a new park. Each plant sold for $6. The club sold 417 plants. How much money did the club raise?

$2,502

23. Marla is paid $4 per hour for baby-sitting. Last month, Marla baby-sat for 22 hr. How much money did she earn?

$88.00

Name ______________________

Subtracting Unlike Fractions

Subtract. Write the answer in simplest form.

1. $\frac{1}{2} - \frac{1}{5}$ **$\frac{3}{10}$**
2. $\frac{6}{7} - \frac{1}{4}$ **$\frac{17}{28}$**
3. $\frac{9}{10} - \frac{3}{5}$ **$\frac{3}{10}$**
4. $\frac{7}{8} - \frac{1}{2}$ **$\frac{3}{8}$**
5. $\frac{3}{4} - \frac{5}{8}$ **$\frac{1}{8}$**
6. $\frac{4}{5} - \frac{1}{3}$ **$\frac{7}{15}$**
7. $\frac{5}{8} - \frac{1}{10}$ **$\frac{21}{40}$**
8. $\frac{1}{2} - \frac{1}{6}$ **$\frac{1}{3}$**
9. $\frac{7}{10} - \frac{1}{4}$ **$\frac{9}{20}$**
10. $\frac{5}{6} - \frac{1}{3}$ **$\frac{1}{2}$**
11. $\frac{11}{12} - \frac{1}{4}$ **$\frac{2}{3}$**
12. $\frac{9}{10} - \frac{1}{6}$ **$\frac{11}{15}$**
13. $\frac{3}{4} - \frac{1}{12}$ **$\frac{2}{3}$**
14. $\frac{6}{7} - \frac{1}{3}$ **$\frac{11}{21}$**
15. $\frac{4}{5} - \frac{1}{6}$ **$\frac{19}{30}$**
16. $\frac{3}{4} - \frac{1}{2}$ **$\frac{1}{4}$**
17. $\frac{2}{3} - \frac{3}{8}$ **$\frac{7}{24}$**
18. $\frac{3}{5} - \frac{1}{15}$ **$\frac{8}{15}$**
19. $\frac{13}{14} - \frac{2}{7}$ **$\frac{9}{14}$**
20. $\frac{1}{3} - \frac{1}{5}$ **$\frac{2}{15}$**
21. $\frac{7}{10} - \frac{2}{5}$ **$\frac{3}{10}$**
22. $\frac{4}{7} - \frac{1}{3}$ **$\frac{5}{21}$**
23. $\frac{7}{12} - \frac{1}{4}$ **$\frac{1}{3}$**
24. $\frac{7}{15} - \frac{2}{5}$ **$\frac{1}{15}$**
25. $\frac{2}{5} - \frac{1}{3}$ **$\frac{1}{15}$**
26. $\frac{7}{9} - \frac{1}{3}$ **$\frac{4}{9}$**
27. $\frac{2}{3} - \frac{2}{7}$ **$\frac{8}{21}$**
28. $\frac{5}{8} - \frac{1}{3}$ **$\frac{7}{24}$**
29. $\frac{2}{3} - \frac{1}{9}$ **$\frac{5}{9}$**
30. $\frac{5}{6} - \frac{1}{2}$ **$\frac{1}{3}$**

Mixed Applications

31. Ron picks $\frac{3}{4}$ qt of strawberries. He gives $\frac{1}{3}$ qt to a neighbor. What quantity of strawberries does he have left?

$\frac{5}{12}$ qt

32. Josh hikes $\frac{11}{12}$ mi before noon and $\frac{5}{6}$ mi after noon. How much farther does Josh hike before noon than after?

$\frac{1}{12}$ mi

33. The auditorium is being set up for a concert. Each row has 32 chairs. How many rows will be needed to seat 832 people?

26 rows

34. Tickets to the concert cost $5. A total of 825 tickets were sold. How much money did the orchestra raise?

$4,125

Name ______________________________

Estimating Sums and Differences

Round each fraction. Write *about 0, about $\frac{1}{2}$,* or *about 1.*

1. $\frac{4}{5}$ about 1
2. $\frac{1}{8}$ about 0
3. $\frac{5}{9}$ about $\frac{1}{2}$
4. $\frac{4}{7}$ about $\frac{1}{2}$
5. $\frac{9}{11}$ about 1
6. $\frac{3}{17}$ about 0
7. $\frac{1}{9}$ about 0
8. $\frac{3}{8}$ about $\frac{1}{2}$
9. $\frac{1}{16}$ about 0
10. $\frac{7}{8}$ about 1
11. $\frac{11}{12}$ about 1
12. $\frac{3}{7}$ about $\frac{1}{2}$

Estimate the sum or difference. **Possible estimates are given.**

13. $\frac{4}{5} + \frac{1}{8}$ 1
14. $\frac{2}{3} - \frac{5}{6}$ 0
15. $\frac{1}{10} + \frac{4}{7}$ $\frac{1}{2}$
16. $\frac{3}{4} + \frac{9}{10}$ 2
17. $\frac{1}{2} + \frac{1}{9}$ $\frac{1}{2}$
18. $\frac{11}{12} - \frac{3}{5}$ $\frac{1}{2}$
19. $\frac{7}{12} + \frac{7}{10}$ $1\frac{1}{2}$
20. $\frac{4}{5} - \frac{13}{15}$ 0
21. $\frac{11}{20} + \frac{8}{9}$ $1\frac{1}{2}$
22. $\frac{8}{15} + \frac{13}{14}$ $1\frac{1}{2}$
23. $\frac{1}{7} + \frac{9}{11}$ 1
24. $\frac{4}{7} - \frac{1}{2}$ 0
25. $\frac{5}{11} + \frac{8}{10}$ $1\frac{1}{2}$
26. $\frac{3}{4} - \frac{1}{9}$ 1
27. $\frac{5}{9} + \frac{8}{11}$ $1\frac{1}{2}$
28. $\frac{8}{10} - \frac{3}{5}$ $\frac{1}{2}$
29. $\frac{8}{9} - \frac{7}{8}$ 0
30. $\frac{4}{7} + \frac{1}{12}$ $\frac{1}{2}$
31. $\frac{3}{5} + \frac{6}{7}$ $1\frac{1}{2}$
32. $\frac{13}{14} - \frac{9}{10}$ 0

Mixed Applications

33. The school chorus is made up of $\frac{4}{9}$ sixth graders and $\frac{11}{12}$ seventh graders. About how many more seventh graders than sixth graders are in the chorus?

about $\frac{1}{2}$ more seventh graders

34. About $\frac{3}{5}$ of the cupcakes for a bake sale are yellow, and $\frac{1}{10}$ of the cupcakes are chocolate. About how many more of the cupcakes are yellow rather than chocolate?

about $\frac{1}{2}$ more yellow

35. Rose buys $\frac{13}{16}$ yd of fabric. She uses $\frac{3}{8}$ yd for a school project. How much fabric does she have left?

$\frac{7}{16}$ yd

36. Max has $\frac{7}{8}$ c sugar. He uses $\frac{1}{3}$ c to make muffins. How much sugar does Max have left?

$\frac{13}{24}$ c

Use with text pages 104–107.

Name ____________________

LESSON 6.1

Adding Mixed Numbers

Write the addition problem shown by the diagram. Then find the sum. Write the answer in simplest form.

1. $1\frac{1}{5} + 1\frac{2}{5} = 2\frac{3}{5}$

2. $2\frac{1}{2} + 1\frac{1}{8} = 3\frac{5}{8}$

3. 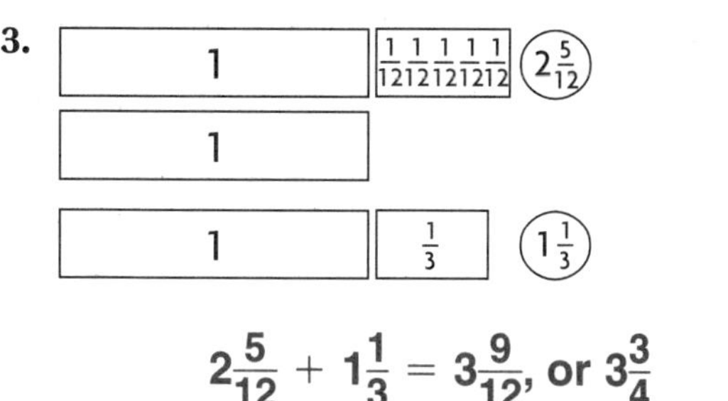

$2\frac{5}{12} + 1\frac{1}{3} = 3\frac{9}{12}$, or $3\frac{3}{4}$

4. 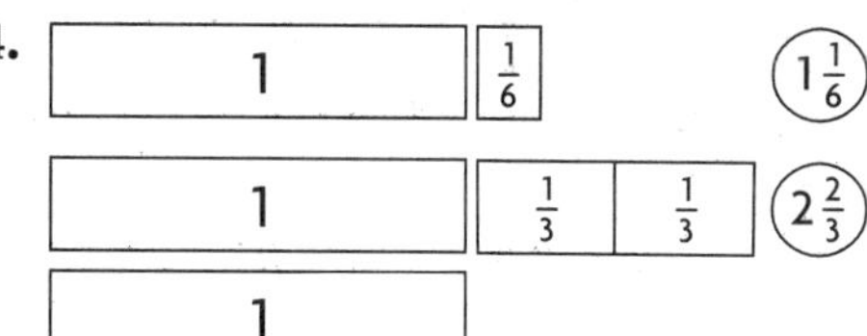

$1\frac{1}{6} + 2\frac{2}{3} = 3\frac{5}{6}$

Draw a diagram to find each sum. Write the answer in simplest form. **Check students' diagrams.**

5. $1\frac{2}{5} + 1\frac{2}{5}$ **$2\frac{4}{5}$**
6. $2\frac{3}{8} + 1\frac{1}{4}$ **$3\frac{5}{8}$**
7. $2\frac{1}{6} + 1\frac{1}{3}$ **$3\frac{3}{6}$, or $3\frac{1}{2}$**
8. $3\frac{1}{2} + 1\frac{1}{4}$ **$4\frac{3}{4}$**
9. $2\frac{3}{8} + 1\frac{1}{2}$ **$3\frac{7}{8}$**
10. $2\frac{2}{3} + 1\frac{1}{6}$ **$3\frac{5}{6}$**

Mixed Applications

For Problems 11–14, use a diagram to solve.

11. Ryan uses $2\frac{1}{3}$ m of wire for his science project. Sean uses $1\frac{1}{6}$ m of wire for his project. How much wire do Ryan and Sean use?

$3\frac{1}{2}$ m

12. The school band practices for $2\frac{1}{2}$ hr on Friday and $3\frac{1}{4}$ hr on Saturday. How many hours does the band practice?

$5\frac{3}{4}$ hr

13. The Garden Club held a bake sale. The club sold 42 cakes for $9.50 each, 188 cookies for $0.20 each and 124 cupcakes for $0.45 each. How much did the club raise?

$492.40

14. The wagot factory makes 7,400 wagots every hour. At this rate, how many wagots are produced during an 8-hour shift?

59,200 wagots

Name ____________________

LESSON 6.2

Subtracting Mixed Numbers

Write each mixed number by renaming one whole.

1. $3\frac{1}{2}$ — $2\frac{3}{2}$
2. $5\frac{3}{4}$ — $4\frac{7}{4}$
3. $2\frac{7}{8}$ — $1\frac{15}{8}$
4. $4\frac{5}{6}$ — $3\frac{11}{6}$
5. $3\frac{5}{12}$ — $2\frac{17}{12}$
6. $4\frac{3}{10}$ — $3\frac{13}{10}$
7. $2\frac{5}{8}$ — $1\frac{13}{8}$
8. $3\frac{2}{9}$ — $2\frac{11}{9}$
9. $5\frac{3}{7}$ — $4\frac{10}{7}$
10. $4\frac{3}{11}$ — $3\frac{14}{11}$

Draw a diagram to find each difference. Write the answer in simplest form. **Check students' diagrams.**

11. $4\frac{4}{5} - 2\frac{2}{5}$ — $2\frac{2}{5}$
12. $3\frac{1}{3} - 1\frac{2}{3}$ — $1\frac{2}{3}$
13. $6\frac{2}{5} - 4\frac{3}{5}$ — $1\frac{4}{5}$
14. $6\frac{5}{8} - 2\frac{1}{8}$ — $4\frac{1}{2}$
15. $3\frac{1}{2} - 2\frac{1}{4}$ — $1\frac{1}{4}$
16. $5\frac{3}{4} - 1\frac{1}{8}$ — $4\frac{5}{8}$
17. $5\frac{1}{6} - 3\frac{2}{3}$ — $1\frac{1}{2}$
18. $4\frac{3}{5} - 2\frac{1}{10}$ — $2\frac{1}{2}$
19. $4\frac{3}{4} - 3\frac{1}{2}$ — $1\frac{1}{4}$
20. $4\frac{1}{8} - 2\frac{3}{4}$ — $1\frac{3}{8}$
21. $5\frac{1}{9} - 3\frac{2}{3}$ — $1\frac{4}{9}$
22. $3\frac{1}{2} - 1\frac{4}{5}$ — $1\frac{7}{10}$

Mixed Applications

For Problems 23–26, draw a diagram to solve. Write the answer in simplest form. **Check students' diagrams.**

23. Kate has $5\frac{5}{8}$ m of wire. She uses $2\frac{1}{8}$ m for a school project. How much wire is left?

$3\frac{1}{2}$ m

24. The school cafeteria has $5\frac{1}{4}$ qt of grape juice and $2\frac{3}{8}$ qt of apple juice. How much more grape juice than apple juice is there?

$2\frac{7}{8}$ qt

25. Michele read for $2\frac{1}{4}$ hr on Monday and $1\frac{1}{2}$ hr on Tuesday. How much time did she spend reading?

$3\frac{3}{4}$ hr

26. Carey works at a local theater during the summer. She is paid $4.75 per hour and works 6 hr each day. If she works 4 days each week, how much does she earn each week?

$114

Use with text pages 116–117.

Name ____________________

Adding and Subtracting Mixed Numbers

Rename the fractions using the LCD. Rewrite the problem.

1. $1\frac{1}{3} + 2\frac{5}{6}$ — $1\frac{2}{6} + 2\frac{5}{6}$
2. $2\frac{3}{4} + 1\frac{1}{2}$ — $2\frac{3}{4} + 1\frac{2}{4}$
3. $5\frac{5}{6} + 3\frac{1}{4}$ — $5\frac{10}{12} + 3\frac{3}{12}$
4. $6\frac{1}{5} + 2\frac{3}{4}$ — $6\frac{4}{20} + 2\frac{15}{20}$

Rename the fraction as a mixed number. Write the new mixed number.

5. $5\frac{10}{7}$ — $1\frac{3}{7}$; $6\frac{3}{7}$
6. $6\frac{7}{4}$ — $1\frac{3}{4}$; $7\frac{3}{4}$
7. $8\frac{9}{5}$ — $1\frac{4}{5}$; $9\frac{4}{5}$
8. $5\frac{13}{10}$ — $1\frac{3}{10}$; $6\frac{3}{10}$

Tell whether you must rename to subtract. Write *yes* or *no*.

9. $8\frac{3}{4} - 6\frac{1}{2}$ — Yes
10. $4\frac{1}{5} - 2\frac{7}{10}$ — Yes
11. $7\frac{1}{4} - 2\frac{2}{3}$ — Yes
12. $5\frac{2}{9} - 3\frac{2}{3}$ — Yes

Find the sum. Write the answer in simplest form.

13. $3\frac{1}{5} + 2\frac{3}{10}$ — $5\frac{1}{2}$
14. $1\frac{3}{8} + 4\frac{1}{2}$ — $5\frac{7}{8}$
15. $6\frac{1}{3} + 2\frac{3}{4}$ — $9\frac{1}{12}$
16. $1\frac{7}{9} + 1\frac{2}{3}$ — $3\frac{4}{9}$

Find the difference. Write the answer in simplest form.

17. $4\frac{2}{3} - 1\frac{1}{2}$ — $3\frac{1}{6}$
18. $5\frac{4}{5} - 3\frac{1}{4}$ — $2\frac{11}{20}$
19. $3\frac{1}{3} - 1\frac{4}{9}$ — $1\frac{8}{9}$
20. $4\frac{5}{8} - 2\frac{1}{2}$ — $2\frac{1}{8}$

Mixed Applications

21. Carol bought a share of stock for $22\frac{1}{8}$ dollars. A week later the stock increased $1\frac{3}{4}$ dollars. What is her share of stock worth now?

    $23\frac{7}{8}$ dollars

22. Bob practices the piano $6\frac{1}{4}$ hr each week. This week he has practiced $4\frac{2}{5}$ hr. How many more hours will Bob practice this week?

    $1\frac{17}{20}$ hr

23. At the Olympics, a souvenir vendor makes $185.50 in sales in the morning and $290.75 in the afternoon. How much money in all does the vendor make that day?

    $476.25

24. A newspaper company surveyed 1,000 people about their major health concern. The major concern of 656 people was keeping fit. How many people did not list keeping fit as a major concern?

    344 people

Name ______________________

Estimating Sums and Differences

Estimate the sum or difference. **Possible estimates are given.**

1. $5\frac{3}{5} + 1\frac{7}{8}$ **$7\frac{1}{2}$**
2. $6\frac{10}{11} - 3\frac{1}{15}$ **4**
3. $4\frac{2}{9} + 3\frac{6}{7}$ **8**
4. $8\frac{7}{9} - 3\frac{11}{12}$ **5**
5. $7\frac{1}{10} + 4\frac{1}{2}$ **$11\frac{1}{2}$**
6. $7\frac{9}{20} - 4\frac{3}{5}$ **3**
7. $5\frac{2}{21} + 1\frac{9}{10}$ **7**
8. $6\frac{4}{7} - 1\frac{1}{13}$ **$5\frac{1}{2}$**
9. $4\frac{5}{8} + 3\frac{4}{5}$ **$8\frac{1}{2}$**
10. $9\frac{11}{14} - 1\frac{7}{10}$ **$8\frac{1}{2}$**
11. $5\frac{2}{3} + 4\frac{1}{6}$ **10**
12. $8\frac{5}{7} - 3\frac{4}{5}$ **$4\frac{1}{2}$**
13. $6\frac{7}{16} - 4\frac{5}{6}$ **$1\frac{1}{2}$**
14. $3\frac{1}{2} + 5\frac{4}{9}$ **9**
15. $7\frac{3}{20} - 2\frac{5}{11}$ **$4\frac{1}{2}$**
16. $10\frac{1}{8} - 6\frac{4}{9}$ **$3\frac{1}{2}$**
17. $9\frac{3}{5} + 3\frac{13}{14}$ **$13\frac{1}{2}$**
18. $6\frac{2}{15} + 4\frac{3}{7}$ **$10\frac{1}{2}$**
19. $7\frac{3}{4} - 1\frac{1}{12}$ **7**
20. $12\frac{11}{21} - 4\frac{3}{8}$ **8**
21. $13\frac{8}{9} - 1\frac{4}{5}$ **12**

Mixed Applications **Possible estimates are given for Exercises 28–29.**

22. Jim has $23\frac{3}{5}$ m of plastic piping. He uses $16\frac{1}{2}$ m on a project. About how much piping does he have left?

about 7 m

23. Joel is riding on a trail that is $6\frac{9}{10}$ mi long. He has traveled $3\frac{2}{5}$ mi. About how much farther does Joel have to ride?

about $3\frac{1}{2}$ m

24. Kyle buys 8 hamburger rolls for $1.28. What is the cost of each roll?

$0.16

25. Sam grew $2\frac{1}{5}$ in. during fifth grade and $1\frac{7}{8}$ in. during sixth grade. How much did Sam grow during the two school years?

$4\frac{3}{40}$ in.

Use with text pages 122–123.

Name ______________________

Multiplying with Fractions

Make a model to find the product.

1. $\frac{1}{2} \times 6$ **3**
2. $\frac{2}{5} \times \frac{1}{2}$ **$\frac{1}{5}$**
3. $\frac{1}{8} \times \frac{1}{2}$ **$\frac{1}{16}$**
4. $10 \times \frac{1}{2}$ **5**
5. $\frac{1}{2} \times \frac{1}{3}$ **$\frac{1}{6}$**

Find the product. Write it in simplest form.

6. $\frac{1}{4} \times \frac{1}{6}$ **$\frac{1}{24}$**
7. $\frac{1}{5} \times \frac{1}{2}$ **$\frac{1}{10}$**
8. $\frac{3}{8} \times \frac{1}{4}$ **$\frac{3}{32}$**
9. $\frac{3}{5} \times \frac{1}{4}$ **$\frac{3}{20}$**
10. $\frac{4}{5} \times \frac{1}{2}$ **$\frac{2}{5}$**
11. $\frac{1}{4} \times \frac{8}{9}$ **$\frac{2}{9}$**
12. $\frac{3}{4} \times \frac{2}{7}$ **$\frac{3}{14}$**
13. $\frac{5}{9} \times \frac{9}{10}$ **$\frac{1}{2}$**
14. $\frac{5}{6} \times \frac{2}{5}$ **$\frac{1}{3}$**
15. $\frac{6}{7} \times \frac{2}{3}$ **$\frac{4}{7}$**

Complete the multiplication sentence.

16. $\frac{5}{6} \times \frac{\boxed{1}}{3} = \frac{5}{18}$
17. $\frac{4}{5} \times \frac{2}{\boxed{7}} = \frac{8}{35}$
18. $\frac{2}{3} \times \frac{\boxed{4}}{7} = \frac{8}{21}$
19. $\frac{9}{\boxed{10}} \times \frac{10}{11} = \frac{9}{11}$
20. $\frac{3}{4} \times \frac{\boxed{1}}{2} = \frac{3}{8}$
21. $\frac{5}{13} \times \frac{2}{\boxed{5}} = \frac{2}{13}$

Mixed Applications

22. Sally knows that $\frac{2}{5}$ of the students in her class are in the school band. Of those students, $\frac{3}{4}$ own their instruments. What fraction of Sally's class own their own instruments?

 $\frac{3}{10}$

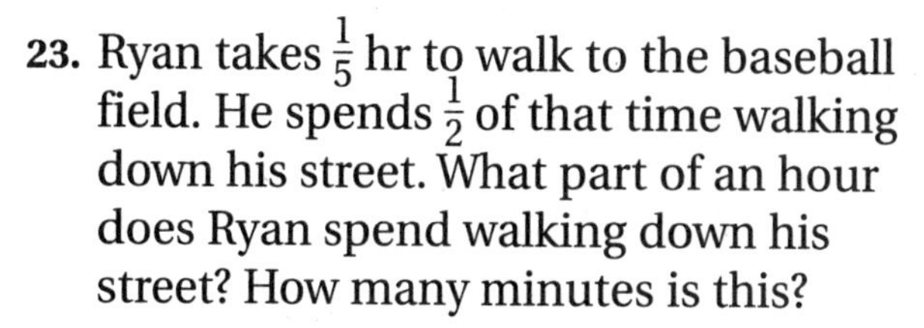

23. Ryan takes $\frac{1}{5}$ hr to walk to the baseball field. He spends $\frac{1}{2}$ of that time walking down his street. What part of an hour does Ryan spend walking down his street? How many minutes is this?

 $\frac{1}{10}$ hour; 6 minutes

24. Bill chose a number, added 2, multiplied the sum by 4 and divided by 8. The final number was 4. What number had Bill chosen?

 6

25. A tour bus travels 19.5 mi each day taking people on sightseeing trips. How many miles does the tour bus travel during a 6-day week?

 117 mi

Name ______________________________

Simplifying Factors

Tell each GCF you would use to rewrite the fractions.

1. $\frac{3}{4}, \frac{4}{5}$ **4**
2. $\frac{8}{9}, \frac{11}{16}$ **8**
3. $\frac{7}{10}, \frac{5}{21}$ **5, 7**
4. $\frac{9}{11}, \frac{2}{3}$ **3**
5. $\frac{6}{25}, \frac{5}{18}$ **6, 5**
6. $\frac{12}{13}, \frac{1}{30}$ **6**
7. $\frac{4}{15}, \frac{10}{17}$ **5**
8. $\frac{7}{9}, \frac{27}{28}$ **7, 9**

Use GCFs to simplify the factors. Write the new problem.

9. $\frac{3}{4} \times \frac{8}{9}$ $\frac{1}{1} \times \frac{2}{3}$
10. $\frac{3}{4} \times \frac{8}{15}$ $\frac{1}{1} \times \frac{2}{5}$
11. $\frac{1}{6} \times \frac{8}{9}$ $\frac{1}{3} \times \frac{4}{9}$
12. $\frac{3}{8} \times \frac{4}{12}$ $\frac{1}{2} \times \frac{1}{4}$
13. $\frac{2}{8} \times \frac{4}{5}$ $\frac{2}{2} \times \frac{1}{5}$
14. $\frac{3}{8} \times \frac{4}{12}$ $\frac{1}{2} \times \frac{1}{4}$
15. $\frac{3}{5} \times \frac{10}{15}$ $\frac{1}{1} \times \frac{2}{5}$
16. $\frac{4}{5} \times \frac{20}{36}$ $\frac{1}{1} \times \frac{4}{9}$

Use GCFs to simplify the factors so that the answer is in simplest form.

17. $\frac{5}{6} \times \frac{3}{10}$ $\frac{1}{4}$
18. $\frac{7}{10} \times \frac{2}{5}$ $\frac{7}{25}$
19. $\frac{1}{2} \times \frac{12}{13}$ $\frac{6}{13}$
20. $18 \times \frac{2}{6}$ **6**
21. $\frac{4}{5} \times 30$ **24**
22. $\frac{9}{10} \times \frac{2}{3}$ $\frac{3}{5}$
23. $\frac{7}{8} \times 24$ **21**
24. $\frac{9}{11} \times \frac{22}{27}$ $\frac{2}{3}$

Mixed Applications

25. The boys' locker room is being remodeled. On Monday, $\frac{2}{9}$ of the lockers were replaced. On Tuesday, $\frac{4}{9}$ of the lockers were replaced. How many of the 504 lockers have been replaced?

336 lockers

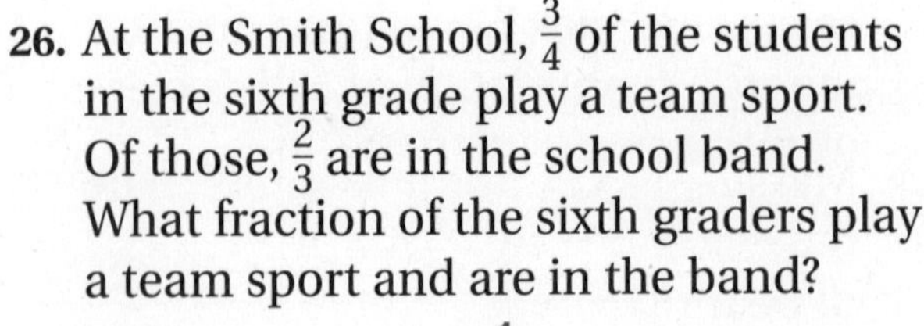

26. At the Smith School, $\frac{3}{4}$ of the students in the sixth grade play a team sport. Of those, $\frac{2}{3}$ are in the school band. What fraction of the sixth graders play a team sport and are in the band?

$\frac{1}{2}$

27. Jon chose a number, added 6, multiplied by 4, and subtracted 2. The result was 70. What's the number?

12

28. On Monday the band rehearsed for $2\frac{1}{4}$ hr. On Wednesday they rehearsed for $3\frac{1}{2}$ hr. How many hours in all did the band rehearse?

$5\frac{3}{4}$ **hr**

Use with text pages 132–133.

Name ______________________

LESSON 7.3

Mixed Numbers

Rewrite the problem by changing each mixed number to a fraction.

1. $3\frac{1}{2} \times \frac{6}{14}$ — $\frac{7}{2} \times \frac{6}{14}$
2. $\frac{3}{8} \times 2\frac{2}{11}$ — $\frac{3}{8} \times \frac{24}{11}$
3. $6\frac{2}{5} \times 1\frac{2}{8}$ — $\frac{32}{5} \times \frac{10}{8}$
4. $2\frac{1}{2} \times \frac{4}{10}$ — $\frac{5}{2} \times \frac{4}{10}$
5. $2\frac{1}{7} \times 2\frac{4}{5}$ — $\frac{15}{7} \times \frac{14}{5}$
6. $1\frac{2}{8} \times \frac{2}{5}$ — $\frac{10}{8} \times \frac{2}{5}$
7. $\frac{8}{9} \times 2\frac{1}{4}$ — $\frac{8}{9} \times \frac{9}{4}$
8. $1\frac{3}{10} \times \frac{15}{26}$ — $\frac{13}{10} \times \frac{15}{26}$

Find the product. Write it in simplest form.

9. $2\frac{1}{2} \times 1\frac{1}{3}$ — $3\frac{1}{3}$
10. $3\frac{1}{5} \times 2\frac{1}{2}$ — 8
11. $8\frac{3}{4} \times \frac{2}{5}$ — $3\frac{1}{2}$
12. $3\frac{1}{3} \times 1\frac{1}{5}$ — 4
13. $3\frac{1}{3} \times 2\frac{2}{5}$ — 8
14. $1\frac{3}{4} \times \frac{3}{14}$ — $\frac{3}{8}$
15. $4\frac{2}{5} \times \frac{10}{11}$ — 4
16. $\frac{6}{7} \times 2\frac{1}{10}$ — $1\frac{4}{5}$
17. $3\frac{1}{2} \times 1\frac{1}{4}$ — $4\frac{3}{8}$
18. $2\frac{3}{5} \times 1\frac{2}{3}$ — $4\frac{1}{3}$
19. $4\frac{3}{8} \times \frac{1}{2}$ — $2\frac{3}{16}$
20. $6\frac{4}{5} \times \frac{5}{8}$ — $4\frac{1}{4}$
21. $2\frac{1}{4} \times 3\frac{1}{5}$ — $5\frac{1}{5}$
22. $9\frac{1}{3} \times 1\frac{2}{7}$ — 12
23. $\frac{3}{5} \times 1\frac{2}{3}$ — 1
24. $12\frac{1}{3} \times 1\frac{1}{2}$ — $18\frac{1}{2}$
25. $1\frac{1}{8} \times \frac{1}{3}$ — $\frac{3}{8}$
26. $3\frac{3}{4} \times 1\frac{5}{6}$ — $6\frac{7}{8}$
27. $2\frac{2}{5} \times 1\frac{5}{8}$ — $2\frac{3}{5}$
28. $5\frac{3}{5} \times 1\frac{2}{7}$ — $7\frac{1}{5}$

Mixed Applications

29. Marsha had $2\frac{2}{3}$ c of mixed nuts. The nuts are $\frac{1}{4}$ peanuts. How many cups of peanuts does she have?

    $\frac{2}{3}$ c

30. Suzanne is practicing for the track meet. She runs $2\frac{1}{2}$ times around the track. A lap of the track is $\frac{1}{4}$ mi. How many miles does she run?

    $\frac{3}{8}$ mi

31. In March, it rained a total of 5.3 in. The normal rainfall for March is 3.9 in. How many inches above normal was the rainfall?

    1.4 in.

32. Ty bought a 7.5 lb roast. He wants it to be medium well. The directions say it should roast for 25 min per lb. How long should he roast the meat?

    187.5 mi or $3\frac{1}{8}$ hr

Use with text pages 134–135.

Name ______________________

Dividing Fractions

Vocabulary

Complete.

1. When you exchange the numerator and denominator of a fraction, this new number is called a **reciprocal** of the fraction.

Find the quotient. Write it in simplest form.

2. $\frac{4}{5} \div \frac{8}{15}$ **$1\frac{1}{2}$**
3. $\frac{7}{10} \div \frac{1}{2}$ **$1\frac{2}{5}$**
4. $\frac{5}{6} \div \frac{1}{2}$ **$1\frac{2}{3}$**
5. $24 \div \frac{1}{2}$ **48**
6. $9 \div \frac{1}{6}$ **54**
7. $\frac{7}{9} \div \frac{2}{3}$ **$1\frac{1}{6}$**
8. $\frac{9}{10} \div \frac{2}{5}$ **$2\frac{1}{4}$**
9. $\frac{9}{20} \div \frac{3}{4}$ **$\frac{3}{5}$**
10. $\frac{5}{8} \div \frac{5}{16}$ **2**
11. $\frac{5}{6} \div \frac{2}{3}$ **$1\frac{1}{4}$**
12. $\frac{12}{21} \div \frac{4}{7}$ **1**
13. $\frac{5}{8} \div \frac{1}{4}$ **$2\frac{1}{2}$**
14. $\frac{3}{4} \div \frac{2}{3}$ **$1\frac{1}{8}$**
15. $\frac{5}{9} \div \frac{5}{6}$ **$\frac{2}{3}$**
16. $\frac{7}{8} \div 12$ **$\frac{1}{16}$**
17. $15 \div \frac{5}{9}$ **27**
18. $\frac{5}{12} \div \frac{3}{4}$ **$\frac{5}{9}$**
19. $\frac{3}{8} \div 18$ **$\frac{1}{48}$**
20. $\frac{7}{10} \div 14$ **$\frac{1}{20}$**
21. $24 \div \frac{4}{5}$ **30**

Mixed Applications

22. In a $\frac{1}{2}$-mi relay, each runner runs $\frac{1}{8}$ mi. How many runners are on the team?

    4 runners

23. Zoe used $\frac{1}{3}$ yd of ribbon to make a bow. How many bows can she make from $4\frac{1}{3}$ yd of fabric?

    13 bows

24. On Saturday, 124 customers went into the Sandwich Shop. Of those customers, $\frac{1}{4}$ ordered the lunch special. How many customers ordered the lunch special?

    31 customers

25. Jeffrey rode his bicycle 1.6 mi to a friend's house. Then they rode 2.5 mi to a park. Then Jeffrey rode 3.8 mi home. How far did Jeffrey ride altogether?

    7.9 mi

Use with text pages 138–139.

Name ______________________________

Work Backward by Dividing Mixed Numbers

Work backward to solve.

1. Seth is trying to decrease the amount of time he spends watching television. This week he spent $4\frac{1}{2}$ hr watching television, which is $\frac{2}{3}$ of the time he watched television last week. How much time did he watch television last week?

   $6\frac{3}{4}$ hr

2. Rich is driving 15 hr to visit his son. He will stop every $2\frac{1}{2}$ hr to rest. How many times will he stop during the trip?

   6 times

3. Sasha wants to plant a rectangular garden. She wants it to have an area of $22\frac{1}{2}$ sq ft. The length is 5 ft. How wide will the garden be?

   $4\frac{1}{2}$ ft

4. Rose has 15 yd of fabric. This is 9 times the amount she needs to make a pillow. How much fabric does she need for each pillow?

   $1\frac{2}{3}$ yd

Mixed Applications

Solve.

CHOOSE A STRATEGY

- Work Backward
- Draw a Diagram
- Use a Formula
- Guess and Check
- Write an Equation
- Make a Table

Choices of strategies will vary.

5. Sam left home and drove to the gym. Then he drove $1\frac{1}{8}$ mi to the bank and $3\frac{3}{4}$ mi to his son's school. Sam drove a total of 9 mi. What is the distance from Sam's home to the gym?

   $4\frac{1}{8}$mi

6. Hakeem has $2.35 in nickels and quarters. He has 23 coins altogether. How many coins of each kind does he have?

   17 nickels and 6 quarters

7. Larry chose a number and multiplied by $2\frac{3}{4}$. The product was 33. What was the number?

   12

8. Carla has a $2\frac{1}{2}$-lb container of plant food. She uses $\frac{1}{8}$ lb every day. How many days will her supply last?

   20 days

9. The area of a rectangular rug is 164 sq ft. The width of the rug is $10\frac{1}{4}$ ft. What is the length of the rug?

   16 ft

10. Rebecca's paper route takes her 10 blocks north of her house, then 5 blocks east, then 2 south, then 3 west, and finally 8 blocks south. How many blocks is Rebecca from her home when she finishes her route?

    2 blocks

Use with text pages 140–141.

Name ______________________

LESSON 8.1

Points, Lines, and Planes

Vocabulary

Write the correct letter from Column 2.

Column 1

d 1. It is an exact location.

c 2. It is a straight path that goes on forever in opposite directions.

a 3. It is part of a line. It has two endpoints.

e 4. It is part of a line. It begins at its endpoint and goes on forever in only one direction.

b 5. It is a flat surface that goes on forever in all directions.

Column 2

a. line segment
b. plane
c. line
d. point
e. ray

For Exercises 6–8, use the figure at the right. **All possible answers are given.**

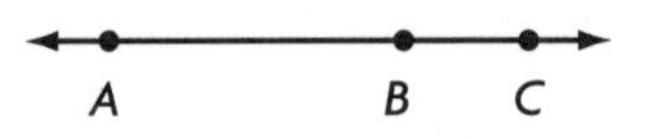

6. Name three points.

point A, point B, point C

7. Name six different rays.

$\overrightarrow{AB}$, $\overrightarrow{BA}$, $\overrightarrow{AC}$, $\overrightarrow{CA}$, $\overrightarrow{BC}$, $\overrightarrow{CB}$

8. Name six different line segments.

$\overline{AB}$, $\overline{BA}$, $\overline{AC}$, $\overline{CA}$, $\overline{BC}$, $\overline{CB}$

Mixed Applications

9. Draw a plan showing the floor of your classroom. Show where you sit, and show where the door is. On your plan, what geometric figure is suggested by the floor of your classroom? by the location of the door?

plane; line segment

10. In a map of your town, suppose you mark the point where you live and the point where your school is. What geometric figure is suggested if you use a ruler to connect the two points?

line segment

11. Mike is grilling soy burgers for a family party. Each burger is $\frac{1}{5}$ lb. How many soy burgers can he make from 6 lb of soy mixture?

30 soy burgers

12. Rita walked $1\frac{1}{3}$ mi to her friend's house. She took a shortcut home, walking $1\frac{1}{4}$ mi. How far did she walk in all?

$2\frac{7}{12}$ mi

Use with text pages 150–151.

Name ______________________

Classifying Lines

Vocabulary

Complete.

1. ____Intersecting____ lines are lines that cross at exactly one point.

2. ____Perpendicular____ lines are lines that intersect to form 90° angles, or right angles.

3. ____Parallel____ lines are lines in a plane that are always the same distance apart. They never intersect and have no common points.

The figure at the right shows 12 lines drawn on the edges of a rectangular prism. **All possible answers are given.**

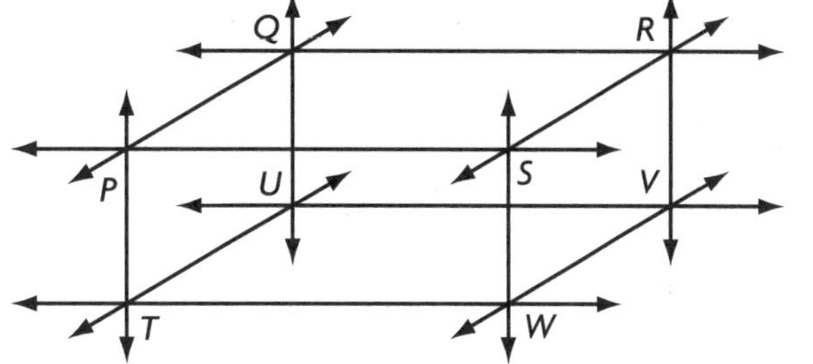

4. Name all lines that are parallel to $\overleftrightarrow{TW}$.

   $\overleftrightarrow{UV}$, $\overleftrightarrow{PS}$, $\overleftrightarrow{QR}$

5. Name all the lines that intersect $\overleftrightarrow{PQ}$.

   $\overleftrightarrow{PS}$, $\overleftrightarrow{QR}$, $\overleftrightarrow{PT}$, $\overleftrightarrow{QU}$

6. Name all the lines that are perpendicular to and intersect $\overleftrightarrow{UQ}$.

   $\overleftrightarrow{PQ}$, $\overleftrightarrow{QR}$, $\overleftrightarrow{TU}$, $\overleftrightarrow{UV}$

Write *true* or *false*. Change any false statement into a true statement.

7. Intersecting lines can be parallel.

   False; parallel lines never intersect.

8. Parallel lines are never perpendicular.

   true

Mixed Applications

9. A ladder leans against a wall. Do the ladder and wall suggest parallel, intersecting, or perpendicular lines?

   intersecting lines

10. At Cheese City, sandwiches are made with $2\frac{1}{2}$ oz of cheese. At House of Cheese, they use $1\frac{3}{4}$ oz. How much more cheese do they put in a sandwich at Cheese City?

   $\frac{3}{4}$ oz

Use with text pages 152–153.

Name ____________________

LESSON 8.3

Angles

Vocabulary

Complete.

1. The measure of a **right** angle is 90°.
2. The measure of a **straight** angle is 180°.
3. The measure of an **obtuse** angle is more than 90° and less than 180°.
4. The measure of an **acute** angle is less than 90°.
5. The **vertex** of an angle is formed by two rays with a common endpoint.

Measure the angle. Write the measure and *acute, right, obtuse,* or *straight.*

6. 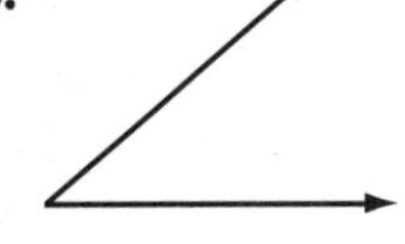

40°, acute

7. 

125°, obtuse

8. 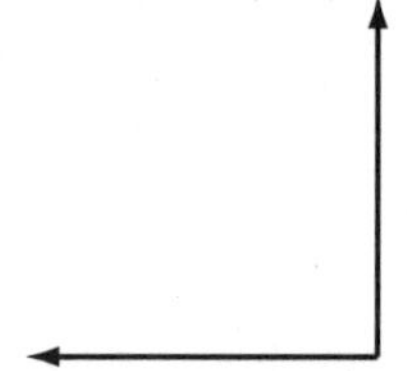

90°, right

9. 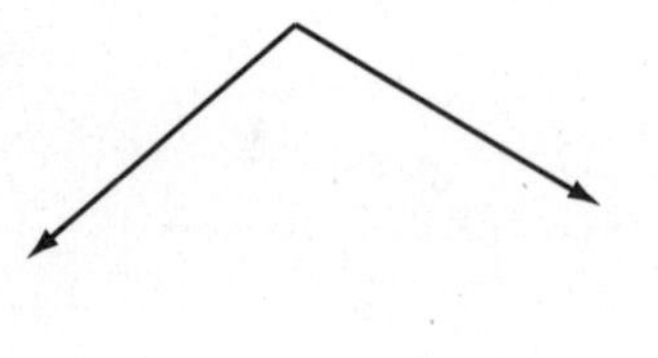

110°, obtuse

10. 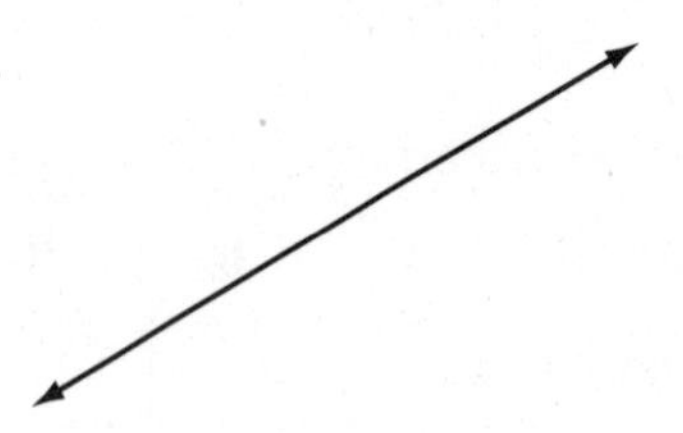

180°, straight

11. 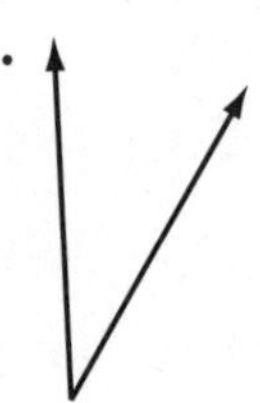

33°, acute

Mixed Applications

12. The hands of a clock indicate that the time is about 4:15. Do the hands form an angle that is obtuse, right, or acute?

acute

13. Mario can wash a car in about $\frac{1}{3}$ hr. How many cars can he wash in $1\frac{2}{3}$ hr?

5 cars

Use with text pages 154–155.

Name ______________________

LESSON 8.4

Constructing Congruent Segments and Angles

Vocabulary

Complete.

1. All line segments of the same length are said to be

 congruent .

2. Two angles that have the same measure in degrees are

 congruent angles.

Use a protractor to measure the angles in each pair. Tell whether they are congruent. Write each measure and *yes* or *no*.

3. 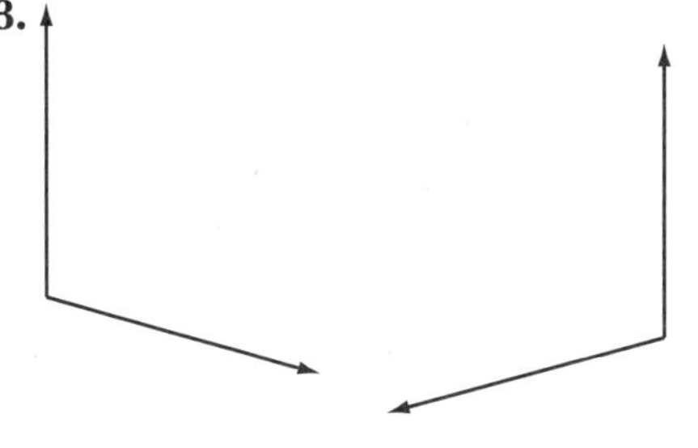

105°, yes

4. 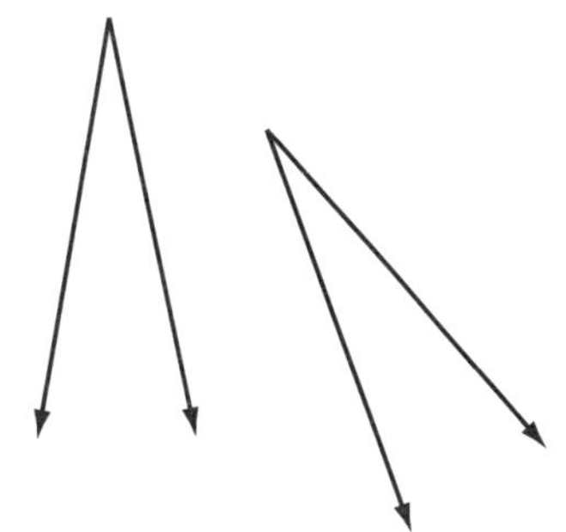

22°, yes

5. 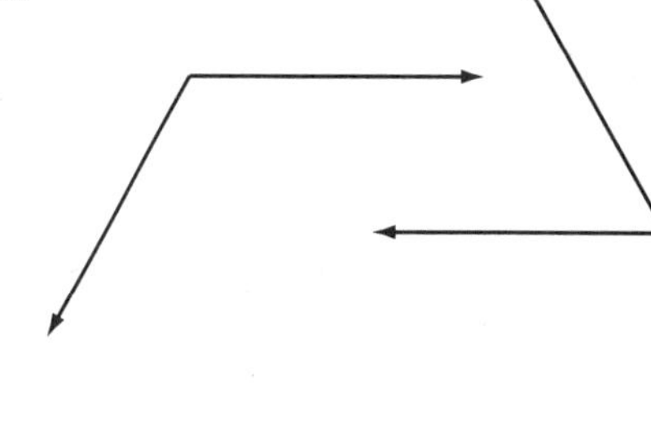

120°, 60°, no

6. In the space at the right, use a compass and a straightedge to construct an angle congruent to the one below.
 Check students' drawings.

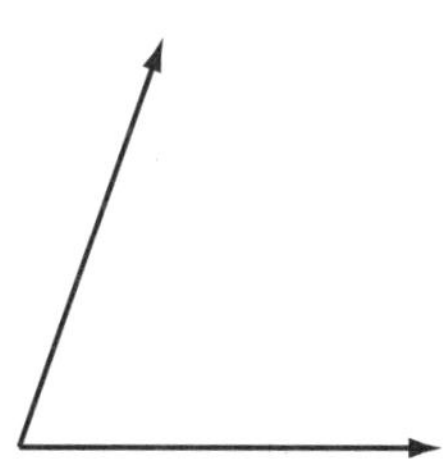

Mixed Applications

7. Look at the four sides of an ordinary window pane. What do you notice about them?

 Opposite sides are congruent.

8. In a science experiment on seeds, Claudia measured the sprout from her seed to be 0.25 in. Paul measured the sprout from his seed as $\frac{1}{8}$ in. Whose seed had sprouted more? Explain.

 Claudia's; $0.25 > \frac{1}{8}$

Use with text pages 156–159.

Name ______________________________

LESSON 8.5

Polygons

Vocabulary

Complete.

1. A ___polygon___ is a closed plane figure formed by three or more line segments.

2. A ___regular___ polygon has all sides congruent and all angles congruent.

Name the quadrilateral. Tell whether both pairs of opposite sides are parallel. Write *yes* or *no*.

3. 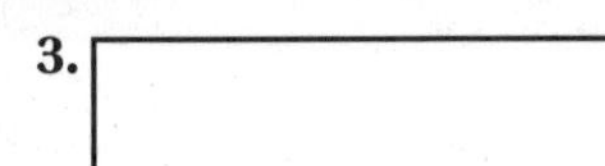

___rectangle; yes___

4. ___parallelogram; yes___

5. 

___trapezoid; no___

For Exercises 6–9, use Figures 1–4. All measurements are in centimeters.

Figure 1

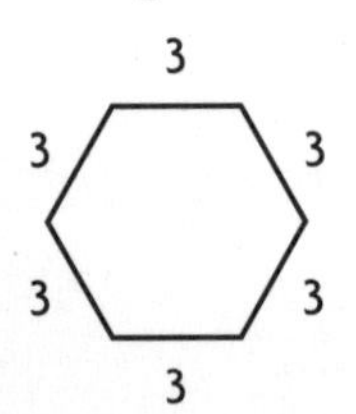

Figure 2

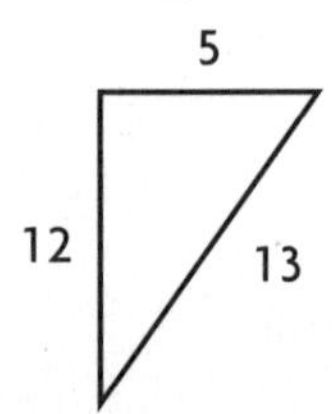

Figure 3

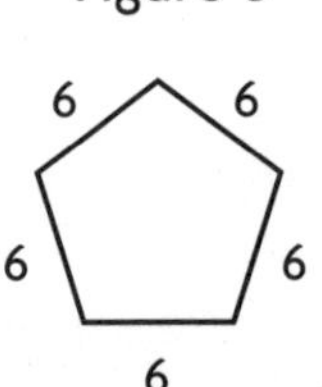

Figure 4

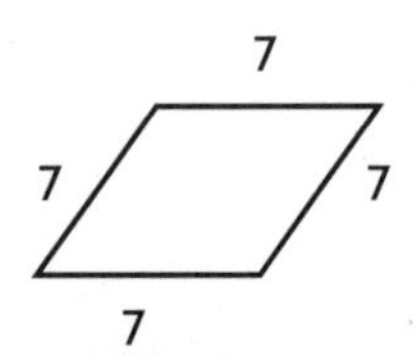

6. Which figure is a pentagon?

___Figure 3___

7. Is Figure 4 a regular polygon?

___no___

8. Is Figure 2 an equilateral, scalene, or isosceles triangle?

___scalene triangle___

9. Which figure is a hexagon?

___Figure 1___

Mixed Applications

10. The Pentagon is a government building in Washington, D.C. How many sides does the building have?

___5 sides___

11. Draw a rectangle. Then draw a line to divide it into two congruent triangles. What type of triangle is each triangle?

___scalene triangle___

Use with text pages 162–163.

Name ______________________

LESSON 9.1

Symmetry and Congruence

Vocabulary

Write the correct letter from Column 2.

	Column 1	Column 2
c	1. line symmetry	a. when a figure can be rotated around a central point and the resulting figure matches the original
b	2. regular polygon	b. all sides and angles are congruent
a	3. rotational symmetry	c. when a figure can be folded so that the two parts of the figure are congruent

Is the dashed line a line of symmetry? Write *yes* or *no*.

4. 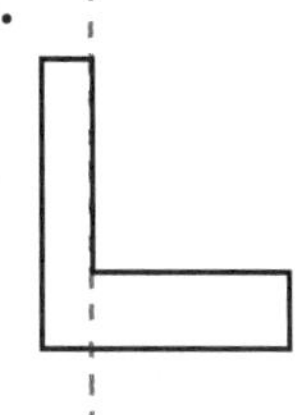

no

5. 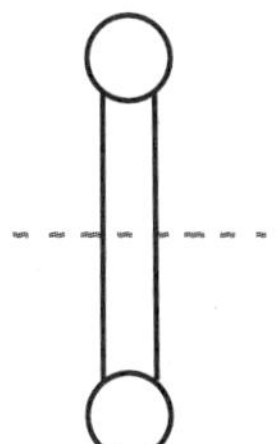

yes

6. 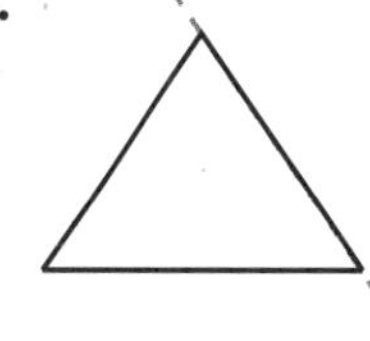

no

Each figure has rotational symmetry. Tell the fraction and the angle measure of each turn.

7. 

$\frac{1}{4}$; 90°

8.  

$\frac{1}{2}$; 180°

9.  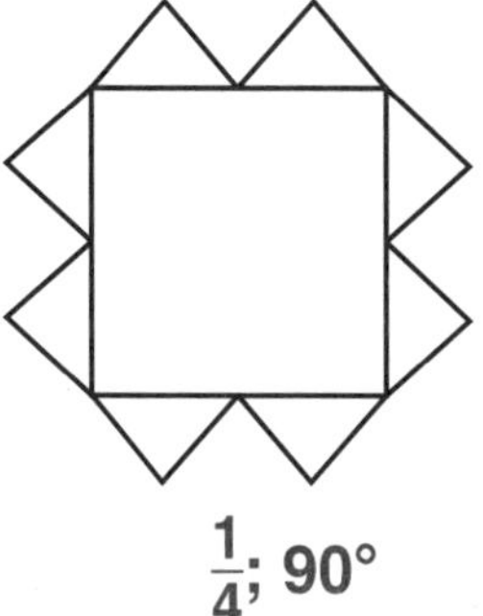

$\frac{1}{4}$; 90°

Mixed Applications

10. Draw a figure that has at least three lines of symmetry.

Check students' drawings.

11. Today Karen has 18 tickets that cost \$10 each. Yesterday she sold \$220 worth of tickets. How many tickets did she start with?

40 tickets

Name ______________________________

LESSON 9.2

Transformations

Vocabulary

Complete each sentence with a term from the Word Box.

Word Box
reflection
rotation
transformation
translation

1. Flipping a figure over a line is called a reflection.
2. A translation is the movement of a figure along a straight line.
3. A movement that doesn't change the size or shape of a figure is called a transformation.
4. Turning a figure around a point is called a rotation.

Tell whether the second figure is a translation or rotation of the first figure.

5. 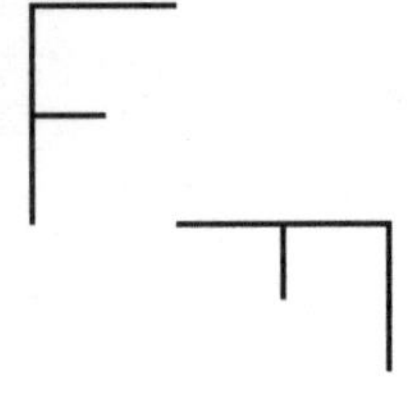

rotation

6. 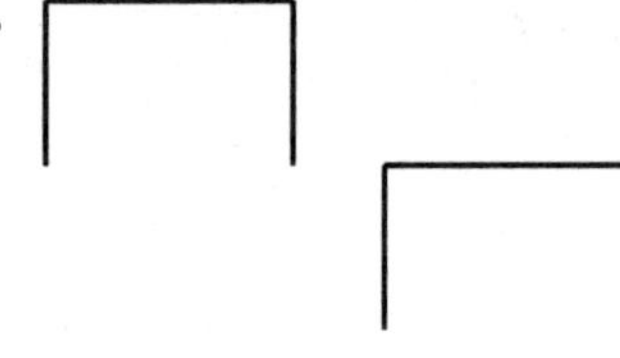

translation

7. 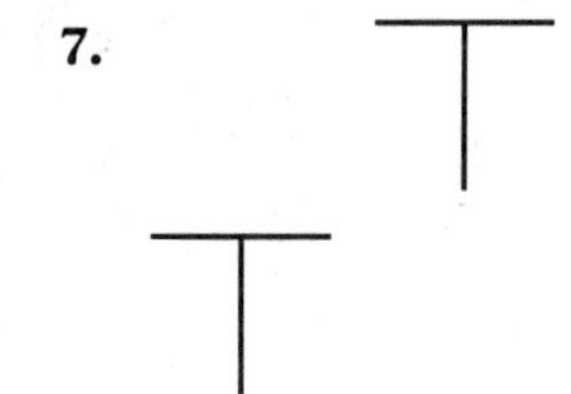

translation

Read the label, and write *true* or *false.* If it is false, name the correct transformation.

8. 

rotation

true

9. 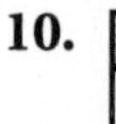

reflection

true

10. 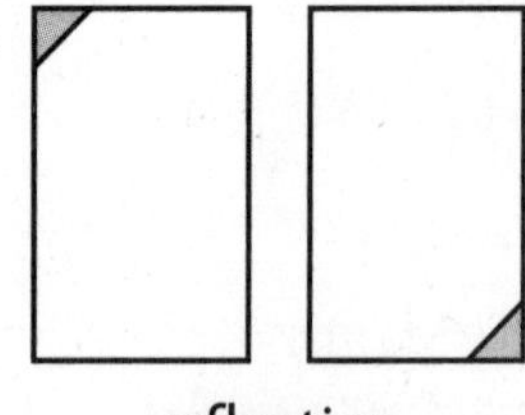

reflection

false; rotation

Mixed Applications

11. Draw a figure. Then draw a reflection of the figure.

Check students' drawings.

12. The sum of two consecutive odd numbers is 28. What are the numbers?

13 and 15

Use with text pages 174–177.

Name ______

LESSON 9.3

Tessellations

Vocabulary

Complete.

1. An arrangement of shapes that completely covers a plane, with no gaps or overlaps, is called a **tessellation**.

Tell whether each shape forms a tessellation. Write *yes* or *no*.

2. 
yes

3. 
yes

4. 
no

5. 
yes

Find the measures of the angles that surround the circled vertex. Then find the sum of the measures.

6. 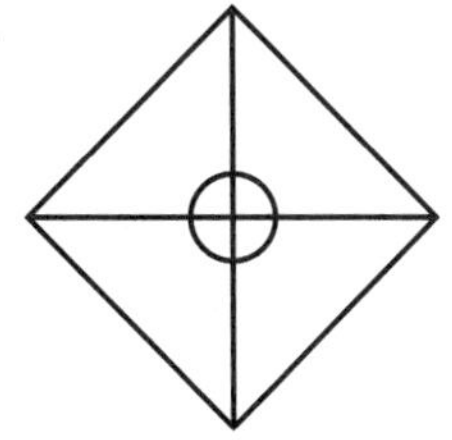
each angle: 90°;
sum: 360°

7. 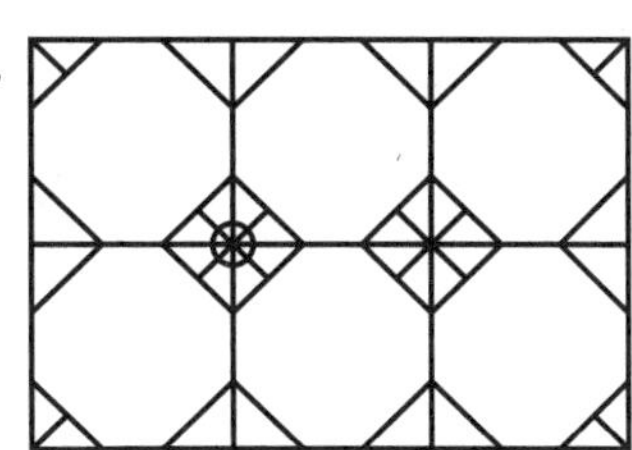
each angle: 45°;
sum: 360°

8. 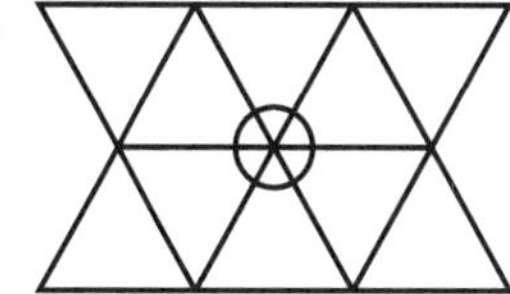
each angle: 60°;
sum: 360°

Mixed Applications

For Problems 9–10, use the figure at the right.

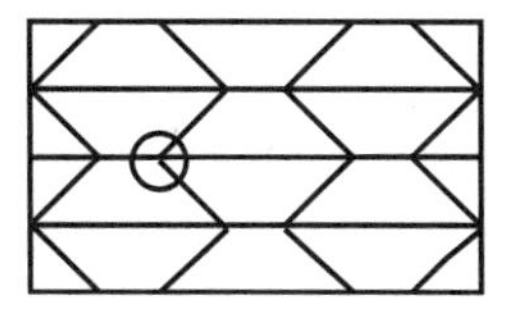

9. What polygons are at the circled vertex?
trapezoids

10. What is the sum of the measures of the angles at the circled vertex?
360°

11. Jon has a square pizza. He cuts it into pieces along all the lines of symmetry of the square. How many pieces does he have?
8 pieces

Name ______________________________

LESSON 9.4

Problem-Solving Strategy

Making a Model

Make a model to solve.

1. Carol is making a design from the shape below. She wants the shape to tessellate a plane. Can she use this shape?

yes

2. Will this shape tessellate a plane? Explain.

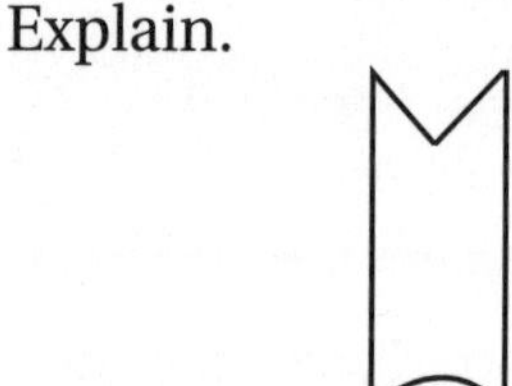

No. The shape does not fit together without gaps.

3. Adam created this shape for a tile floor. Will his shape tessellate a plane?

yes

4. Draw a figure that does NOT tessellate a plane.

Check students' drawings.

Mixed Applications

Solve.

CHOOSE A STRATEGY

- Work Backward • Guess and Check • Write an Equation • Use a Formula • Make a Model • Act It Out

Choices of strategies will vary.

5. Tina bought a television for $375. She made a down payment of $115 and paid the rest in equal payments of $20 a month. How many months did it take her to pay for the television?

13 months

6. Maria is stocking books in the library storeroom. The shelves are 0.8 m apart, and the bottom shelf is 1.5 m from the floor. There are 5 shelves. How far from the floor is the top shelf?

4.7 m

7. Five friends are standing in line for movie tickets. Paula is standing in between Chuck and Chris. Zack is first in line. Chris is in front of Marla. List the order in which they are standing.

Zack, Chuck, Paula, Chris, Marla

8. Paul's budget for food is $10 more than $\frac{1}{3}$ of his total budget. If $100 was allotted for food, what was the total budget?

$270

Use with text pages 180–181.

Name ______________________

LESSON 10.1

Solid Figures

Vocabulary

Complete by using a word from the Word Bank.

Word Bank
base
lateral face
polyhedron
prism
vertex

1. A polyhedron is a solid figure with flat faces that are polygons.

2. A prism is a polyhedron with two congruent and parallel bases.

3. A polyhedron is named by the shape of its base.

4. The point where two or more sides of a solid figure meet is called a vertex.

5. A face that is not a base is called a lateral face.

Name each figure.

6. 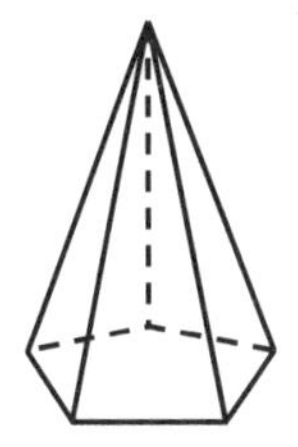

pentagonal pyramid

7. 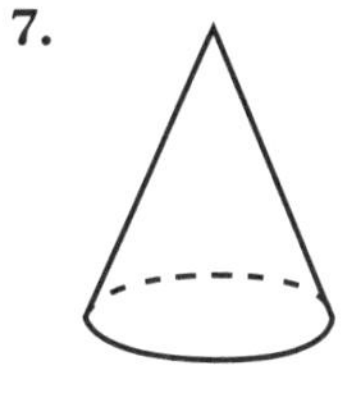

cone

8. 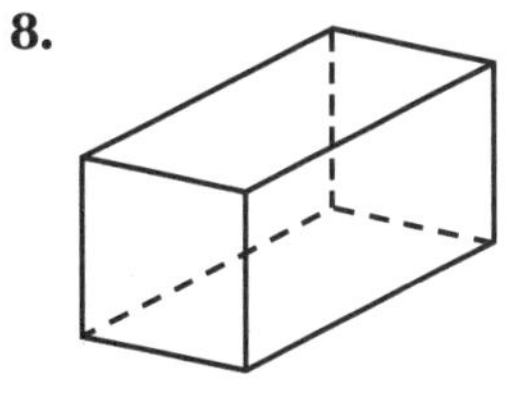

rectangular prism

9. 

cylinder

Write *true* or *false* for each statement. Change each false statement into a true statement.

10. A cylinder has one base.

False; a cylinder has two bases.

11. A cone has one flat surface.

True

Mixed Applications

12. Carole made two pentagonal pyramids and glued the bases together. How many faces does her figure have?

10 faces

13. If $r = 14$, then what is $r - 3$?

11

Name ______________________

Faces, Edges, and Vertices

For Exercises 1–8, use the figure at the right.

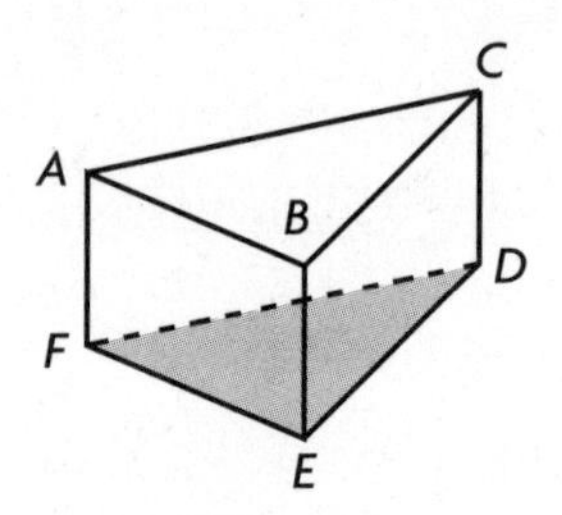

1. Name the vertices.

   A, B, C, D, E, F

2. Name the edges.

   AB, AC, BC, BE, CD, AF, FE, ED, DF

3. Name the faces.

   ABC, DEF, AFEB, BEDC, CDFA

4. Name the figure.

   triangular prism

5. Is the figure a polyhedron? Explain why.

   Yes; it is a solid figure with flat faces that are polygons.

Complete the table.

		Pentagonal Pyramid	Rectangular Prism	Pentagonal Prism	Triangular Pyramid	Triangular Prism	Rectangular Pyramid
6.	Number of edges	10	12	15	6	9	8
7.	Number of faces	6	6	7	4	5	5
8.	Number of vertices	6	8	10	4	6	5

Mixed Applications

9. Scott wants to paint a rectangular prism so no two faces have the same color. What is the least number of colors he needs?

   6 colors

10. A survey showed that $\frac{2}{5}$ of the 160 sixth-grade students play soccer. How many students is that?

    64 students

Use with text pages 188–189.

Name ______________________

LESSON 10.3

Building Solids

Vocabulary

Complete.

1. A __net__ is an arrangement of polygons that folds to form a solid figure.

Use the rectangular prism at the right for Exercises 2–4.

4 in.
2 in.
5 in.

2. What are the dimensions of the faces?

   __2 faces are 4 in. × 5 in.; 2 faces are 2 in. × 5 in.; 2 faces are 2 in. × 4 in.__

3. Draw each of the faces on an index card. **Check students' drawings.**
4. Cut out the faces, and arrange them to form a net for the prism. Tape the pieces together to form the prism. **Check students' prisms.**

Will the arrangement of squares fold to form a cube? Write *yes* or *no*.

5. 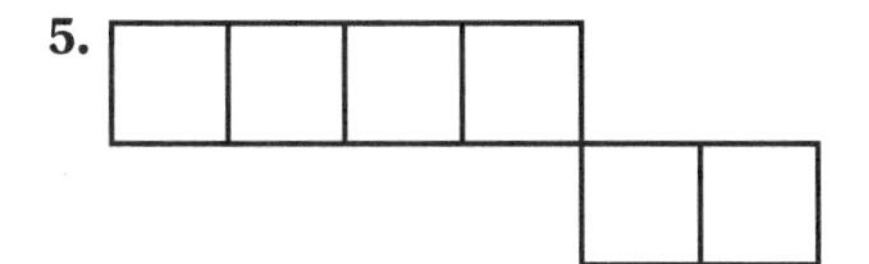

   __No__

6. 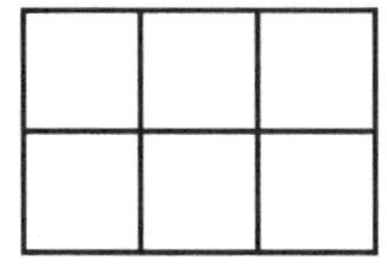

   __No__

7. 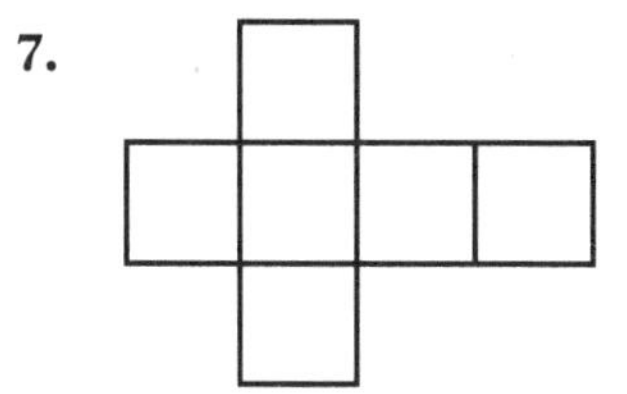

   __Yes__

Mixed Applications

8. Ron has made a net for a prism that is 3 in. high and has two of the triangles shown at the right as the bases. How many faces does his prism have?

   __5 faces__

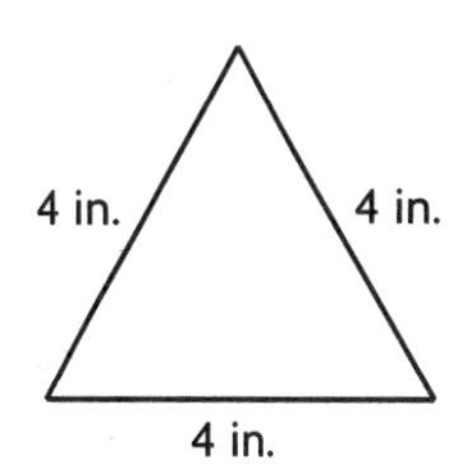

9. Steve has 3 oranges. He cuts each orange into fourths. How many fourths is that?

   __12 fourths__

10. Vanessa uses 1.5 g of cloves in a spice mixture. She makes 5 batches of the mix. How many grams of cloves does Vanessa use altogether?

   __7.5 g of cloves__

Name ______________________________

Two-Dimensional Views of Solids

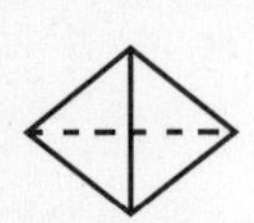
Triangular Pyramid

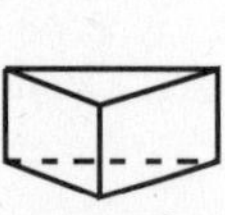
Triangular Prism

Rectangular Pyramid

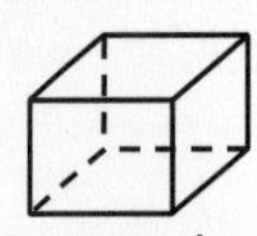
Rectangular Prism

Pentagonal Pyramid

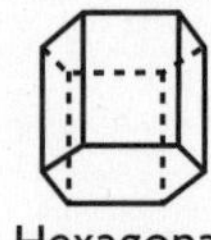
Hexagonal Prism

Cylinder

Cone

Name each solid that has the given top view. Refer to the solids in the box above.

1. 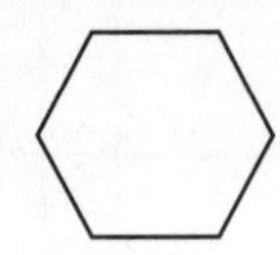

hexagonal prism

2. 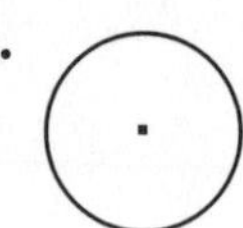

cone

3. 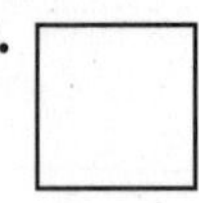

rectangular prism

4. 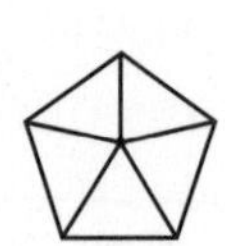

pentagonal pyramid

5. 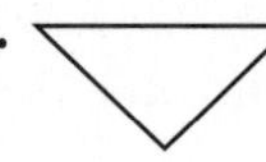

triangular prism

Name the solid figure that has the given views.

6. Top

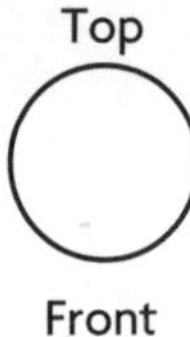

Front

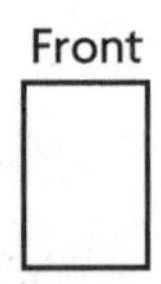

Bottom

cylinder

7. Top

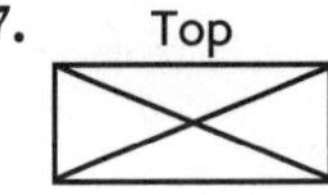

Front

Side

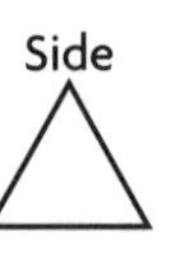

rectangular pyramid

8.

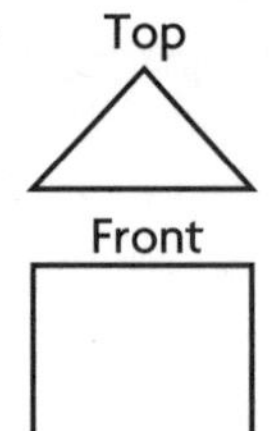

Bottom

triangular prism

For Exercises 9–10, use the solids in the box at the top of the page.

9. Which solid has hexagons in some of its views?

hexagonal prism

10. Which solids have circles in some of their views?

cylinder, cone

Mixed Applications

11. The top and bottom views of a certain solid are triangles. Yet, the front view of the solid is a rectangle. Name the solid.

triangular prism

12. A supermarket receives 12 cases of eggs. There are 24 dozen eggs in each case. How many eggs does the market receive?

3,456 eggs

Use with text pages 194–195.

Name ______________________________

Problem-Solving Strategy

Solve a Simpler Problem

Solve by first solving a simpler problem.

1. Jon is building models of edible prisms. He uses gumdrops for vertices and licorice for edges. How many gumdrops and pieces of licorice will he need to make a prism whose base has 8 sides?

   16 gumdrops and

   24 pieces of licorice

2. Carol wants to make a model of a prism whose base has 9 sides. She will use balls of clay for the vertices and straws for the edges. How many balls of clay and straws will she need? How many faces will her prism have?

   18 balls of clay; 27 straws

   11 faces

3. Chloe used 30 toothpicks as edges to make a model for a prism. How many sides did the base of her prism have? How many vertices?

   10 sides; 20 vertices

4. Dan used 12 balls of clay as vertices to make a model for a prism. How many sides did the base of his prism have? How many edges?

   6 sides; 18 edges

Mixed Applications

Solve.

CHOOSE A STRATEGY

- Work Backward
- Draw a Diagram
- Use a Formula
- Guess and Check
- Write an Equation

Choices of strategies will vary.

5. Alyssa used 8 gal of gasoline to make a 192-mi trip. How much gasoline will she need to make a 360-mi trip?

   15 gal

6. Peter has $4.20 in dimes and quarters. He has 21 coins. How many coins of each kind does he have?

   7 dimes; 14 quarters

7. Zack bought muffins at the bakery. He gave half of the muffins to a friend and divided the rest equally among the six members of his family. If each person in his family got 3 muffins, how many muffins did Zack buy?

   36 muffins

8. Dennis bought 4 books and a bookmark at the bookstore. The bookmark cost $1.25. The total bill was $21.25. What was the average price Dennis paid for each book?

   $5.00

Name ______

LESSON 11.1

Defining the Problem

Solve.

1. Tom's class is in charge of buying decorations for a sixth-grade party. Which of these decisions will they have to make?
   a. the type of food to serve
   b. the type of decorations to buy
   c. the quantity of decorations needed

   b and c

2. What information does the class in Exercise 1 need before they buy decorations?
   a. the size of the room where the party is being held
   b. which beverages to buy
   c. the cost of various types of decorations

   a and c

3. The school band is going to hold a car wash next month. Which of these decisions will they have to make?
   a. the price they should charge to wash a car
   b. the date of their next rehearsal
   c. the supplies they will need to wash cars

   a and c

4. What information does the band in Exercise 3 need before the day of the car wash?
   a. how much people are willing to pay to have a car washed
   b. which store sells car-washing supplies at the lowest price
   c. how many band members own cars

   a and b

5. The Garden Club is going to sell flowers at the high school chorus's spring concert. Which of these decisions will they have to make?
   a. the songs to be sung at the concert
   b. the quantity of flowers to have on hand
   c. the price to charge for the flowers

   b and c

6. Which of the following questions should the Garden Club in Exercise 5 answer before the concert?
   a. How long will the concert be?
   b. How much are people willing to pay for flowers?
   c. How many tickets to the concert have been sold?

   b and c

Mixed Applications

7. The PTO is sponsoring a teacher luncheon. What decisions do they need to make? What information do they need to make the decisions?

   Possible answers: What to serve at the luncheon? How many people will attend? Are cooking/refrigeration facilities available? How much money is available to buy food?

8. Robbie bought 5 boxes of apple juice, 4 boxes of grape juice, 7 boxes of orange juice, and 4 boxes of grapefruit juice. What fraction of the boxes were apple juice?

   $\frac{1}{4}$ of the boxes

Use with text pages 208–209.

Name ______________________

Choosing a Sample

Vocabulary

Write the correct letter from Column 2.

Column 1	Column 2
b 1. population	a. part of a population
c 2. random sample	b. particular group of people
a 3. sample	c. method by which every individual in a given population has an equal chance of being selected

Ron is on a committee to identify the most popular television show among the 280 sixth, seventh, and eighth graders in the school.

4. How many people should he survey?

28 people

5. Whom should he survey?

sixth, seventh, and eighth graders

6. How can Ron get a random sample?

Possible answer: by randomly choosing students at homeroom classes.

7. Is a random sample of only seventh graders fair? Explain.

No; his sample should include sixth and eighth graders also.

Mixed Applications

For Exercises 8–9, use the table at the right.

8. The table shows the results of a survey Susan took at school. How many students did she survey?

110 students

9. Susan surveyed 1 out of every 10 students as they were leaving school. Was this a random sample of the students at the school? Explain.

Yes. Every member of the population had an equal chance of being selected.

FAVORITE SUBJECT	
Subject	Number of Students
Reading	34
Math	27
Social Studies	18
Science	31

10. A sweater is on sale for $\frac{2}{3}$ of its original price. What is the sale price of the sweater, if the original price was $36?

$24

Name ______________________

LESSON 11.3

Bias in Surveys

Vocabulary

Complete.

1. A sample is **biased** if any individual in the population is not represented in the sample.

Tell whether the sampling method is *biased* or *not biased.* Explain.

The Tri-State Soccer League is conducting a survey to determine if the players want to change the style of soccer shirt.

2. Randomly survey all players who wear size large shirts.

   biased; excludes players who wear other sizes

3. Randomly survey all members of championship teams.

   biased; excludes members of non-championship teams

4. Randomly survey 80 players.

   not biased; all have an equal chance

5. Randomly survey all league coaches.

   biased; excludes all players

Determine whether the question is biased.

6. Do you feel that country music is better than all other types of music?

   biased

7. What type of team sport do you enjoy playing?

   not biased

Mixed Applications

For Exercises 7–8, use the following information.

The French club is conducting a survey to see how many students might be interested in learning more about French culture.

8. Describe how a random sample can be chosen.

   Possible answer: A computer can randomly choose 40 members.

9. Describe how a random sample might be biased.

   Possible answer: survey by gender or by team

10. Kyle rode his bicycle a total of 48 km at a rate of 8 km per hr. How long did he ride?

    6 hr

11. Joanne earns $24.50 per hr as a construction worker. How much does she earn if she works 7.5 hr?

    $183.75

 Use with text pages 212–213.

Name ______________________________

Collecting and Organizing Data

Vocabulary

Complete each sentence by writing a term from the Word Bank.

Word Bank
cumulative frequency
frequency table
range
tally table

1. A ___tally table___ is a table that has categories that allow you to record each piece of data as it is collected.

2. A running total of the number of people surveyed is called ___cumulative frequency___.

3. The difference between the greatest number and least number in a set of numbers is the ___range___.

4. A ___frequency table___ gives a total for each category or group in a set of data.

For Exercises 5–6, use the data in the box at the right.

Students' Heights (cm)					
160	129	158	155	136	128
128	159	142	147	148	144
152	133	135	136	162	158
139	160	128	139	159	144
155	147	136	148	162	133

5. Make a line plot.

128	129	133	135	136	139	142	144	147	148	152	155	158	159	160	162
X X X	X	X X	X	X X X	X X	X	X X	X X	X X	X	X X	X X	X X	X X	X X

6. Find the range. ___34___

Mixed Applications

7. Make a line plot for reading test scores of 98, 75, 82, 100, 96, 100, 81, 78, 100, 92, 84, 86, 78, 100, 78. Then determine the range, and make a frequency table that uses intervals. **Check students' tables.**

   ___range = 25___

8. Rick surveyed his classmates about their color preferences. He collected this data: 6 prefer red, 8 prefer blue, 2 prefer yellow, 5 prefer green, and 4 prefer white. Organize this information in a tally table and in a frequency table. **Check students' tables.**

Name ________________________________

LESSON 12.1

Using Graphs to Display Data

Vocabulary

Complete.

1. Stem-and-leaf plots order the data from ____least____ to ____greatest____.

Make an appropriate graph for the set of data. **Check students' graphs.**

2.

Month	May	June	July	Aug	Sept
High Temperature (in °F)	56	71	86	93	79

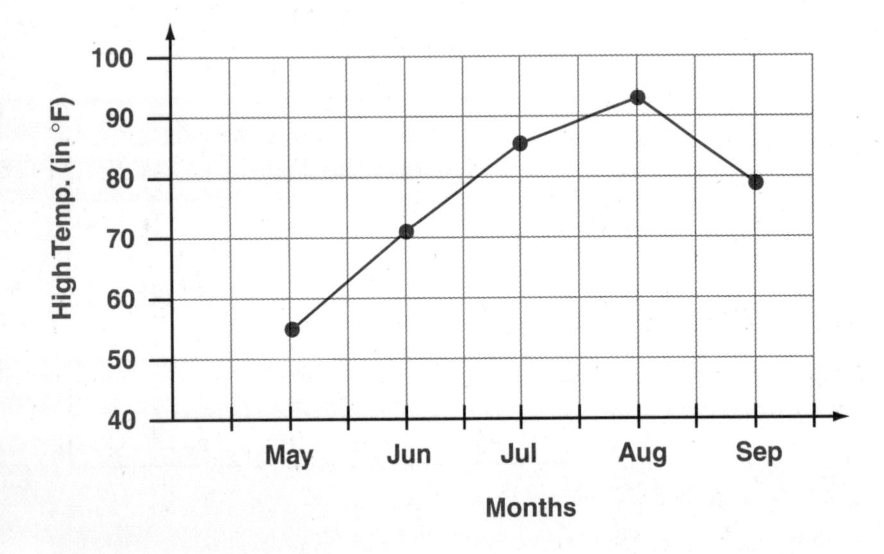

Mixed Applications

3. Make a stem-and-leaf plot for Janet's math test scores: 95, 83, 78, 90, 75, 85, 91, 98, 80.

Janet's Math Test Scores

Stem	Leaves
7	5 8
8	0 3 5
9	0 1 5 8

Key: 8|3 = 83

4. Mrs. Lopez bought her class some pencils for \$0.05 each and erasers for \$0.03 each. She bought 17 items in all and spent \$0.69. How many of each did she buy?

____9 pencils and 8 erasers____

 Use with text pages 224–227.

Name ______________________

LESSON 12.2

Histograms

Vocabulary

Write the letter for the correct graph.

1. bar graph __c__
2. histogram __a__
3. line graph __b__

a. 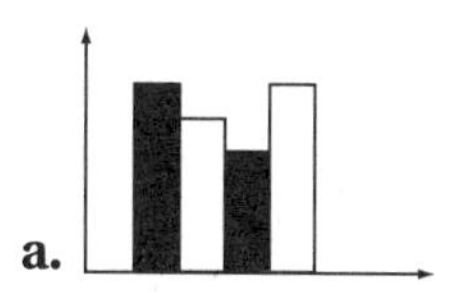

b. 

c. 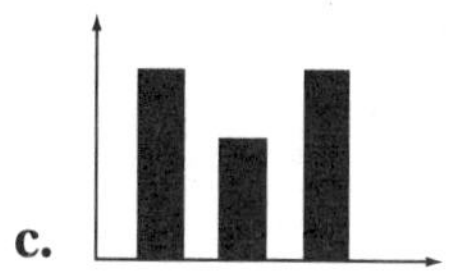

Tell whether a bar graph or a histogram is more appropriate.

4. frequency of fish caught at different times of day

   histogram

5. average monthly telephone bill for a year

   bar graph

6. number of shoppers in a store during 3 different time intervals

   histogram

For Exercises 7–8, use the table below.

Campers at Day Camp				
Age	5–7	8–10	11–13	14–16
Number	6	11	18	9

7. Make a histogram. **Check students' graphs.**

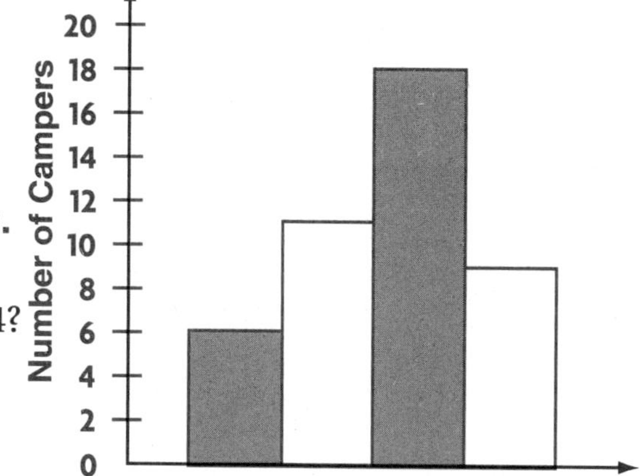

8. How would the number of campers in each group change if you used 5 groups instead of 4?

   number of campers per group would decrease

Mixed Applications

For Exercise 9, use the histogram at the right.

9. During what time period did the most flights arrive?

   9:00–10:59

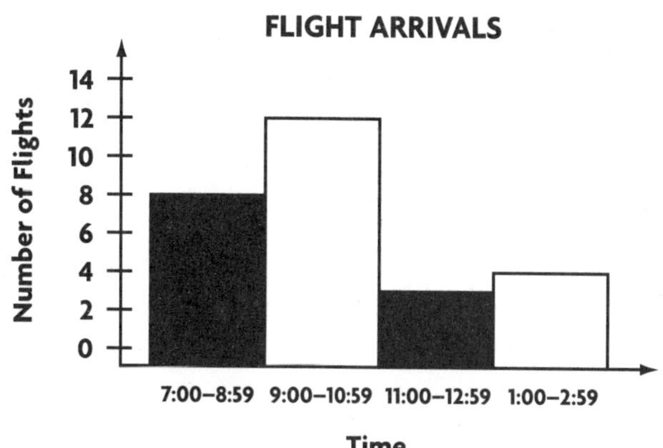

10. Chen baby-sits for $2.50 per hour. If he baby-sat 6 hr last week and 3 hr this week, how much did he make?

    $22.50

Name ______________________________

Graphing Two or More Sets of Data

1. Make a multiple-bar graph of the homework data below. **Check students' graphs.**

HOURS SPENT DOING HOMEWORK

Name	Science	Math
Nigel	2.5 h	0.5 h
Marty	1 h	1.5 h
Julie	0.75 h	2 h
Luis	1.25 h	1 h

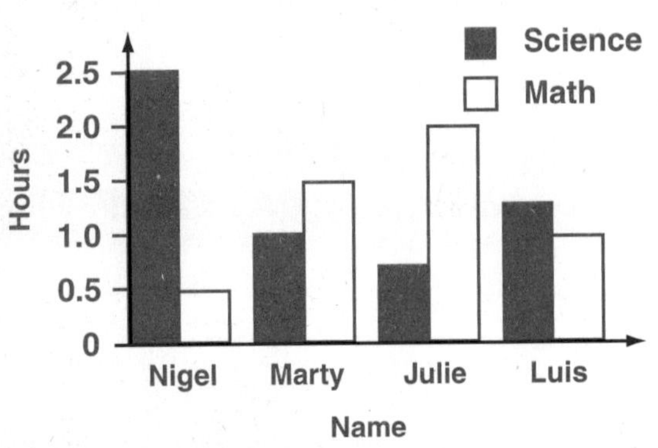

2. Make a multiple-line graph of the temperature data below. **Check students' graphs.**

AVERAGE LOW TEMPERATURE

Year	Jan	Feb	Mar
1994	−4°F	6°F	8°F
1995	6°F	3°F	12°F
1996	−10°F	9°F	16°F
1997	12°F	−5°F	14°F

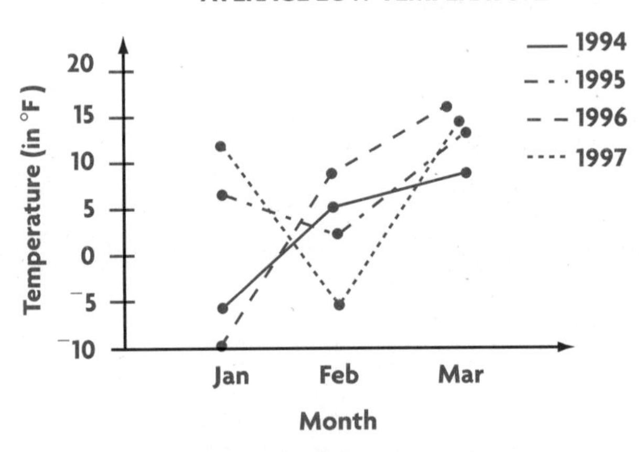

Mixed Applications

Gretchen researched the number of new students who came to her school during 5 months of the school years 1996 and 1997. Her data are shown.

NUMBER OF NEW STUDENTS

	Sept	Oct	Jan	Feb	Mar
1996	35	12	10	6	9
1997	5	23	14	0	12

3. What kind of graph would you use for the data? Explain.

   multiple-bar graph; to compare two sets of data

4. Make a graph for the data. **Check students' graphs.**

5. Bill has 180 baseball cards. He has 3 times as many infielders as outfielders. How many of each does he have?

   135 infielders; 45 outfielders

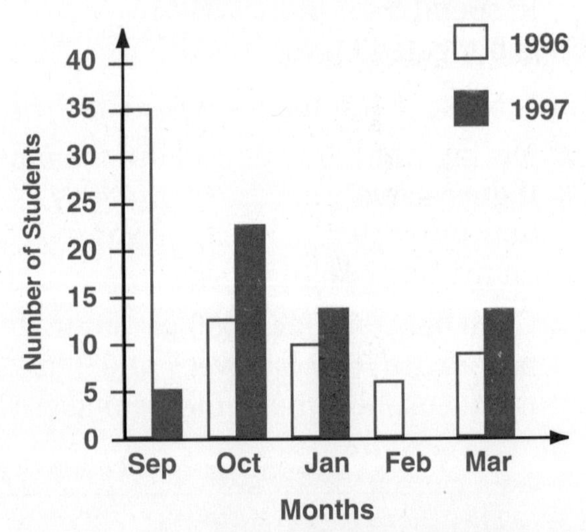

Use with text pages 230–231.

Name ______________________________

LESSON 12.4

Making Circle Graphs

For Exercises 1–2, use the circle graph at the right.

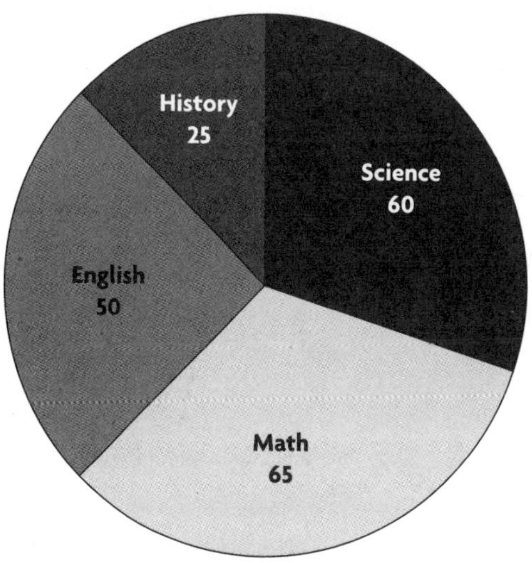

1. How many students were surveyed about their favorite subject?

   200 students

2. What fraction of the circle does Science represent?

   $\frac{60}{200}$, or $\frac{3}{10}$

For Exercises 3–6, use the following data.

The Video Shack kept track of the categories of videos that were rented last weekend.

VIDEOS RENTED			
Drama	Comedy	Action	Classics
250	250	300	200

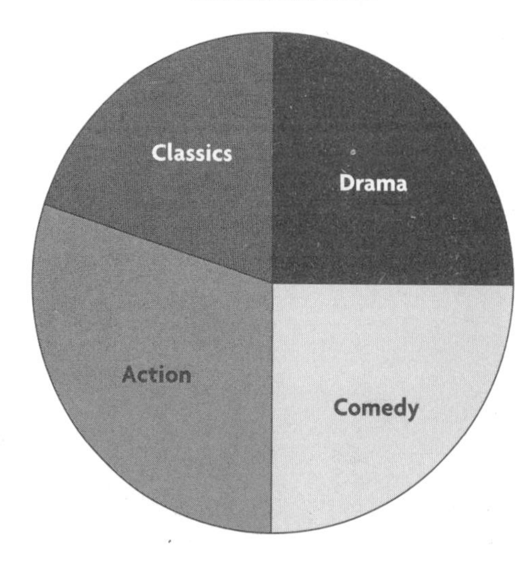

3. Into how many sections would you divide a circle graph for the data?

   4 sections

4. Make a circle graph of the data.

Check students' graphs.

Mixed Applications

For Exercises 5–6, use the circle graph at the right.

5. The circle graph shows how the Millers plan to spend their money. How many times as much is budgeted for food as for transportation?

   4 times as much

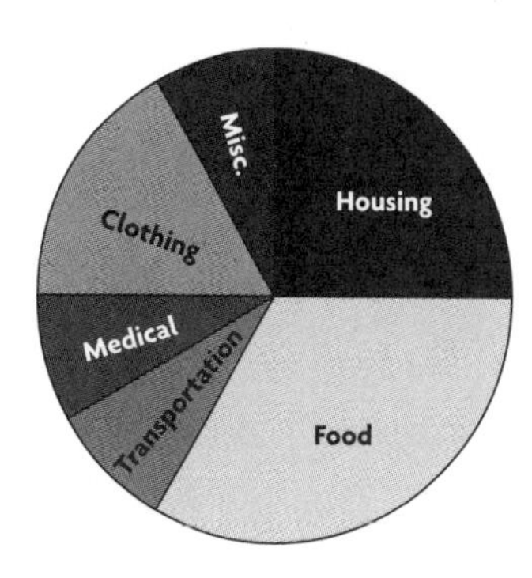

6. If the Millers' total income is $30,000, how much do they plan to spend on medical? on housing?

   $2,500; $7,500

Use with text pages 234–235.

Name ___________________________

Analyzing Graphs

For Exercises 1–3, use the circle graph.

The newspaper's music critic conducted a survey to find out how teens enjoyed a newly released CD. The results are displayed in the circle graph.

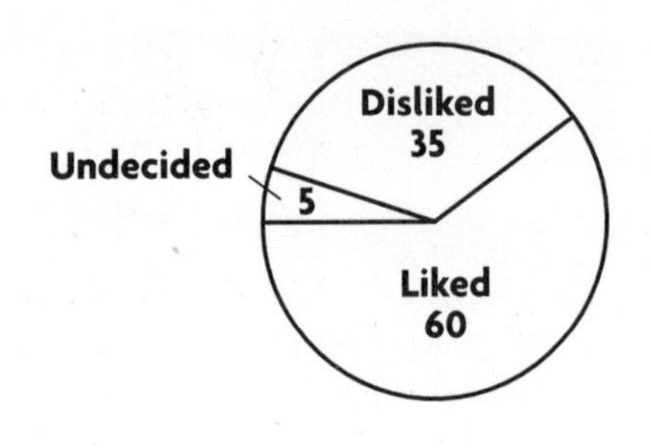

1. Which opinion forms the largest part of the graph?

   liked the CD

2. What did the music critic learn from this survey?

   that the majority of teens liked the CD

3. How many more teens liked the CD than disliked it?

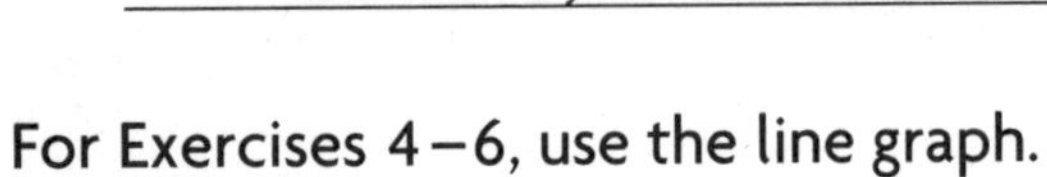

   25 more, or almost twice as many

For Exercises 4–6, use the line graph.

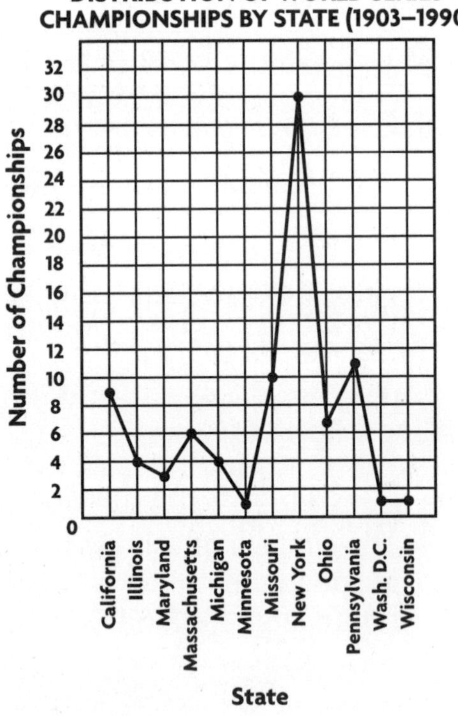

4. Which state had the greatest number of World Series championships?

   New York

5. What is the relationship between the state with the greatest number of championships and the states with the least number of championships?

   NY had 29 more championships than Wisconsin, Minnesota, or Washington, DC.

6. What factors could have caused such a difference in the number of championships won by each state?

   Possible answers: the number of baseball teams in each state; the number of years each team was in existence

Mixed Applications

7. What type of relationship should you look for when analyzing a bar graph?

   Identify the relationships shown by the axes.

8. Tim gave a clerk $9.05 for a book and received $4.00 in change. How much was the book?

   $5.05

Use with text pages 240–241.

Name ______________________

LESSON 13.2

Misleading Graphs

For Exercises 1–4, use the bar graph at the right.

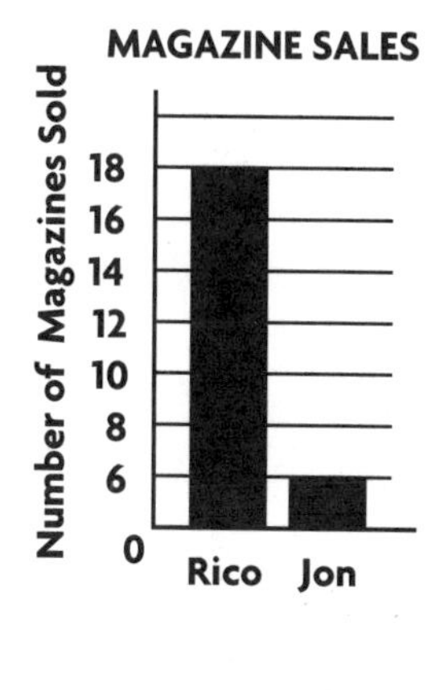

1. Who sold more magazines? **Rico**
2. About how many times as high is the bar for Rico's sales as the bar for Jon's sales? **about seven times**
3. Did Rico sell seven times as many magazines as Jon? Explain.
   No, the uneven interval between 0 and 6 makes the graph misleading.
4. On the back of this sheet, make a new graph, one that is not misleading, to show Rico's and Jon's magazine sales.
   Check students' graphs.

For Exercises 5–8, use the line graphs at the right.

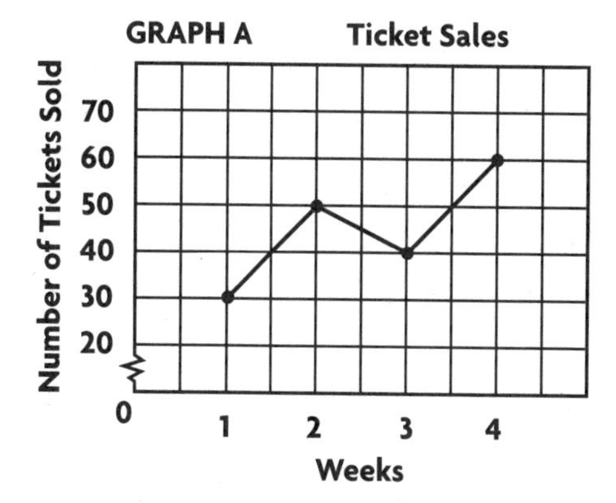

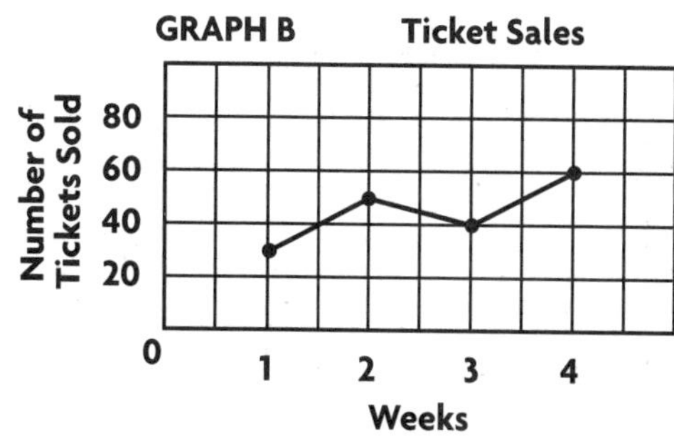

5. In Graph A, about how many times as great as the sales in Week 1 do the sales in Week 4 appear? **about three times as great**
6. In Graph B, about how many times as great as the sales in Week 1 do the sales in Week 4 appear? **about twice as great**
7. What was the actual difference in sales between Week 1 and Week 4? Were the Week 4 sales three times as great? **60 – 30 = 30; no, they were twice as great.**
8. Which graph gives a better picture of the data? Explain. **Graph B; the intervals are the same on the scale.**

Mixed Applications

9. Are the data in a misleading graph factual? Explain.
   Yes, the data are factual. The presentation is misleading.

10. A battery company claims that its 18-month warranty lasts 3 times as long as any other warranty. According to the company, what is the longest of the other warranties?
    6 months

Name ____________________

Making Predictions

For Exercises 1–3, use the graph at the right.

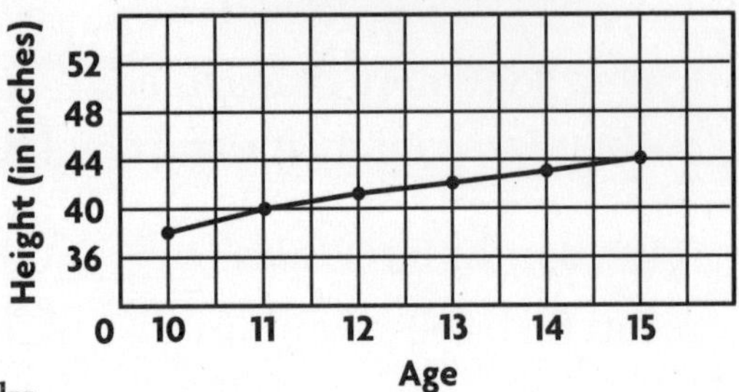

1. How much did Shannan's height change from age 10 to age 11? **2 in.**
2. What has been the pattern for Shannan's growth? **increasing about 1 in. per year**
3. What do you predict Shannan's height will probably be at age 16? **45 in.**

For Exercises 4–7, use the graph at the right.

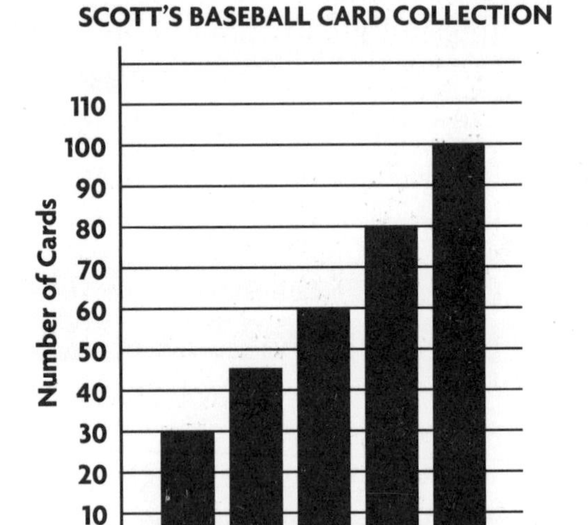

4. How many cards did Scott have in 1992? in 1993? **30; 45**
5. How many more cards did Scott have in 1995 than in 1994? **20 more**
6. What relationship exists between the numbers of cards Scott had in the first and last years shown on the graph? **about 3 times as many cards in 1996 as in 1992**
7. What number of cards would you predict Scott will have in 1997? **120 cards**

Mixed Applications

8. The telephone bills for a pharmacy over the past four months were $220, $209, $197, and $184. What has the trend been for the telephone bills? **decreasing by $11, then $12, and then $13**
9. The price a particular car sold at was $15,300. After one year, the car was worth $12,100. After two years, the car was worth $8,900. What is the trend for the value of the car? **decreasing by 3,200 each year**
10. Beth paid for lunch with a $20 bill. She received $15.03 in change. How much was her lunch? **$4.97**
11. Rick paid for school supplies with two $5 bills. He received $0.86 change. How much were the school supplies? **$9.14**

Use with text pages 244–245.

Name ______________________

Mean, Median, and Mode

Vocabulary

Write the correct letter from Column 2.

Column 1 **Column 2**

__b__ 1. mean — a. number that appears most often in a group of numbers

__c__ 2. median — b. sum of a group of numbers divided by the number of addends

__a__ 3. mode — c. middle number in a group of numbers arranged in numerical order

---

Complete the table.

	Data	Mean	Median	Mode
4.	12, 15, 11, 15, 13, 10, 15	13	13	15
5.	68, 74, 71, 69, 74, 78, 70	72	71	74
6.	7.6, 6.2, 6.0, 6.2, 8.1, 6.7	6.8	6.45	6.2

Write *true* or *false* for each.

7. The mode is always one of the numbers in the data. __true__

8. Every set of data has a mean. __true__

For Exercises 9–10, use the table below.

Test Scores									
98	88	82	91	83	76	98	100	84	90

9. Make a line plot and use it to find the median and the mode.

Check students' plots; median: 89; mode: 98.

10. Make a stem-and-leaf plot and use it to find the median and the mode.

Check students' plots; median: 89; mode: 98.

Mixed Applications

11. Five books have the following prices: \$12.00, \$14.95, \$13.50, \$13.75, and \$13.80. What are the mean and median?

__mean: \$13.60; median: \$13.75__

12. A baseball league had 576 members. There are 16 members on each team. How many teams are in the league?

__36 teams__

Use with text pages 246–249.

Name ____________________

Box-and-Whisker Graphs

For Exercises 1–3, use the box-and-whisker graph below.

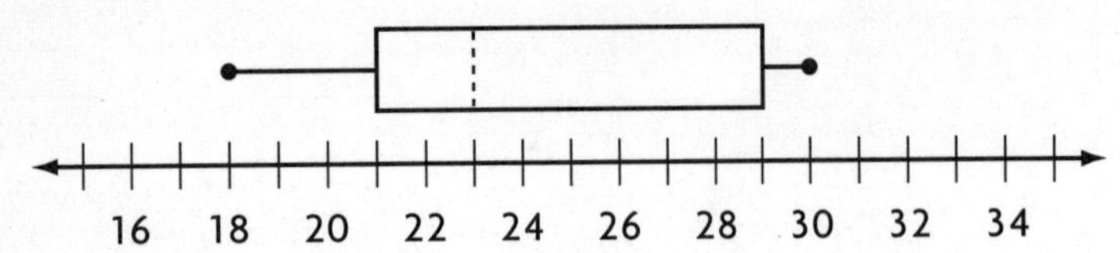

1. What is the median? **23**

2. What are the lower and upper quartiles? **21; 29**

3. What are the lower and upper extremes? **18; 30**

For Exercises 4–12, use the table below.

Lengths of Phone Calls (in min)									
17	21	16	22	24	26	18	28	25	29
21	18	14	23	25	18	26	24	22	23

4. What is the median? **22.5**

5. What is the mode? **18**

6. What is the mean? **22**

7. What is the range? **15**

8. Make a box-and-whisker graph.

   Check students' graphs.

9. What are the lower and upper quartiles? **18; 25**

10. What are the lower and upper extremes? **14; 29**

11. What fractional part of the data are times less than 25 minutes? $\frac{7}{10}$

12. What fractional part of the data are times greater than the mean? $\frac{1}{2}$

Mixed Applications

13. How can you determine the median of a set of data by analyzing a box-and-whisker graph?

    It is represented by a dashed line.

14. The mean of 4 numbers is 21.5. What is the sum of the numbers?

    86

Use with text pages 252–253.

Name ________________________________

LESSON 14.1

Problem-Solving Strategy

Account for All Possibilities

Use the strategy *account for all possibilities* to solve.

1. Tina is making an appointment for a haircut. It can be on Tuesday, Wednesday, or Friday at 11:00 A.M., 1:00 P.M., 2:30 P.M., or 4:00 P.M. Find the total number of choices.

   12 choices

2. Rod is choosing an outfit. He can wear tan, blue, or black pants with a white, green, or print shirt. Find the total number of choices.

   9 choices

3. Larry is buying a new car. He can have a silver, black, white, or blue exterior, with a tan, gray, white, or red interior. How many choices does he have in all?

   16 choices

4. The cafeteria is offering a hamburger, garden salad, turkey sandwich, tuna sub, or pizza, with either milk or juice. How many choices do the students have?

   10 choices

Mixed Applications

Solve.

CHOOSE A STRATEGY

- **Account for All Possibilities**
- **Find a Pattern**
- **Guess and Check**

Choices of strategies will vary.

5. For three weeks, Ted works every fourth day. He starts on Tuesday. What day of the week is his last day?

   Monday

6. On Friday, the theater sold 3 times as many tickets as it sold on Wednesday. On Friday, 225 tickets were sold. How many tickets were sold on Wednesday?

   75 tickets

7. Ann is choosing an outfit. She can wear shorts, a skirt, blue pants, or tan pants with a white sweater, red sweater, or striped blouse. How many choices does she have?

   12 choices

8. Five students are standing in a line. Jon is in front of Terri. Joan is behind Carol. Zack is in front of Jon. Carol is fourth in line. In what order are the students standing?

   Zack, Jon, Terri, Carol, Joan

Name ______________________

LESSON 14.2

Probability

Vocabulary

1. What is probability? Probability is a comparison of the number of favorable outcomes and the number of possible outcomes.

For Exercises 2–5, use the spinner at the right.

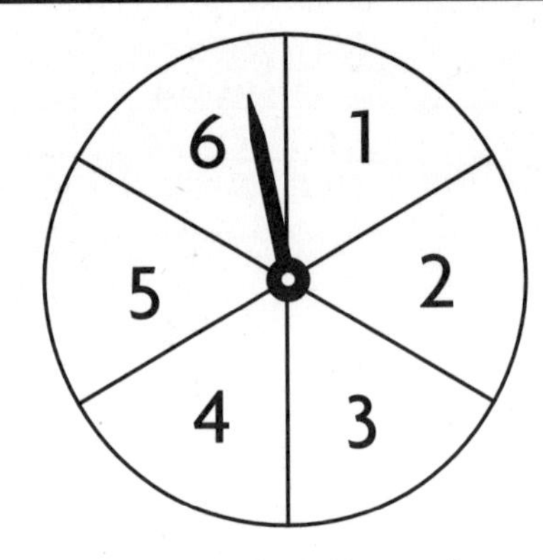

2. How many favorable outcomes are there for choosing 3?

   1

3. How many possible outcomes are there? Name them.

   6; 1, 2, 3, 4, 5, 6

4. What is the probability of choosing 3?

   $\frac{1}{6}$

5. What is the probability of choosing an even number?

   $\frac{3}{6}$, or $\frac{1}{2}$

The letters *H, O, L, I, D, A,* and *Y* are put in a bag. Find each probability.

6. P(*Y*) $\frac{1}{7}$

7. P(*H* or *L*) $\frac{2}{7}$

8. P(*A, I,* or *O*) $\frac{3}{7}$

For Exercises 9–12, use the rectangle at the right. Find each probability.

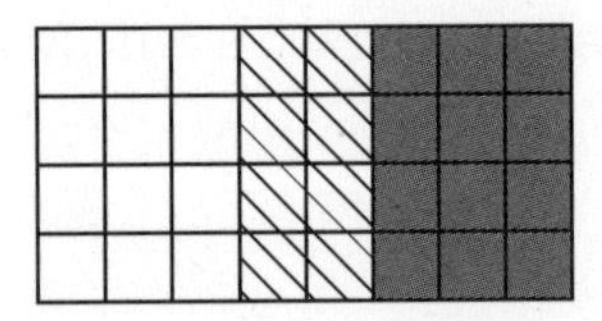

9. P(shaded square) $\frac{3}{8}$

10. P(striped square) $\frac{1}{4}$

11. P(white or shaded square) $\frac{3}{4}$

12. P(white square) $\frac{3}{8}$

Mixed Applications

13. If you have a cube numbered 6 through 11, what is the probability you will roll an odd number?

    $\frac{1}{2}$

14. There are 57 students in an after-school club. There are twice as many boys as girls in the club. How many girls are in the club?

    19 girls

Use with text pages 260–263.

Name ___________________________

More on Probability

For Exercises 1–8, use the spinner at the right. Find each probability on a single spin.

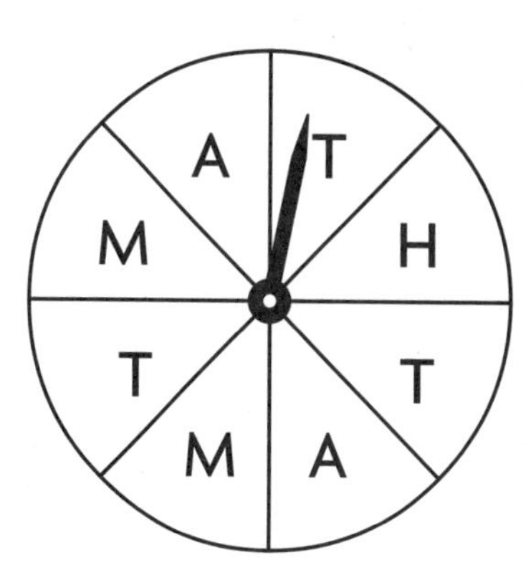

1. P(*M*) $\frac{1}{4}$
2. P(*H*) $\frac{1}{8}$
3. P(*J*) 0
4. P(*T*) $\frac{3}{8}$
5. P(*A*) $\frac{1}{4}$
6. P(*M* or *A*) $\frac{1}{2}$
7. P(*T* or *H*) $\frac{1}{2}$
8. P(*M*, *A*, or *T*) $\frac{7}{8}$

A bag contains 5 blue, 3 red, and 2 green pencils. You choose one pencil without looking. Find each probability.

9. P(red) $\frac{3}{10}$
10. P(blue) $\frac{1}{2}$
11. P(green) $\frac{1}{5}$
12. P(blue or red) $\frac{4}{5}$

Cards numbered 2, 2, 2, 3, 4, 4, 5, and 5 are placed in a bag. You choose one card without looking. Find each probability.

13. P(2) $\frac{3}{8}$
14. P(4 or 5) $\frac{1}{2}$
15. P(3 or 5) $\frac{3}{8}$
16. P(2, 3, or 4) $\frac{3}{4}$

For Exercises 17–19, use the figure at the right. Find each probability.

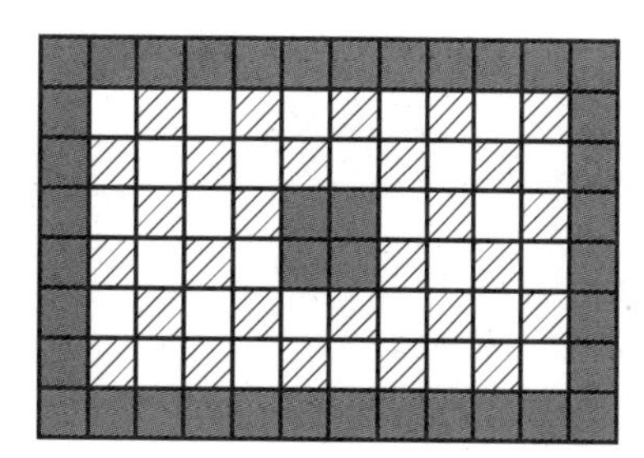

17. P(shaded square) $\frac{5}{12}$
18. P(striped or white square) $\frac{7}{12}$
19. P(shaded or striped square) $\frac{17}{24}$

Mixed Applications

20. There are 8 pairs of socks in Jill's drawer. Three pairs are white, two are blue, two are tan, and one is striped. If Jill chooses a pair without looking, what is the probability of selecting white socks?

    $\frac{3}{8}$

21. Josh wants to get an average bowling score of 135. He will bowl 4 games. How many total points will he need for all 4 games?

    540 points

Name ______________________

LESSON 14.4

Experimental Probability

Vocabulary

1. What is experimental probability?

The experimental probability of an event is the number of times the event occurs compared with the total number of times you do the activity.

Adam tossed a coin 50 times. For Exercises 2–3, use the table at the right to find the experimental probability.

Coin	Heads	Tails
Toss	22	28

2. P(Heads) $\frac{11}{25}$

3. P(Tails) $\frac{14}{25}$

4. What is the mathematical probability of getting heads? $\frac{1}{2}$

Sarah rolled a number cube numbered 1 to 6. The table below shows the results of rolling the cube 50 times. Use the results in the table to find the experimental probability.

Number	1	2	3	4	5	6
Times rolled	6	11	5	10	16	2

5. P(3) $\frac{1}{10}$

6. P(4 or 5) $\frac{13}{25}$

7. P(1 or 2) $\frac{17}{50}$

8. What is the mathematical probability for each number? $\frac{1}{6}$

Mixed Applications

9. Mitch spins the pointer of a spinner 20 times. It lands on blue 15 times. What is the experimental probability of landing on blue?

$\frac{3}{4}$

10. Ryan was hitting in the batting cage. Out of 100 balls, he hit 42 of them. How many balls can he expect to hit in his next 50 tries?

21

11. A car is traveling at a speed of 52 mph. How far does it travel in $3\frac{1}{2}$ hr?

182 mi

12. Scott wants to get a 94 average in science. He will have a total of 9 tests. How many total points will he need for all 9 tests?

846 points

Use with text pages 268–269.

Name ______________________

LESSON 15.1

Numerical and Algebraic Expressions

Vocabulary

Write the correct letter from Column 2.

	Column 1	Column 2
a	1. a mathematical phrase that includes only numbers and operation symbols	a. numerical expression
c	2. an expression that includes a variable	b. variable
b	3. a letter or symbol that stands for one or more numbers	c. algebraic expression

---

Write a numerical or algebraic expression for the word expression.

4. seven less than eleven

   $11 - 7$

5. six more than a number, x

   $x + 6$

6. 8 multiplied by one third

   $8 \times \frac{1}{3}$

7. 84 divided by 8

   $84 \div 8$

Write a word expression for the numerical or algebraic expression. **Possible expressions are given.**

8. $26 \div 3.9$

   26 divided by 3.9

9. $n \times 7.5$

   n times 7.5

10. $15 - r$

    r less than 15

11. $q + 6.3$

    6.3 more than *q*

Mixed Applications

12. Hank has $56. Phillip has $12 less than Hank. Write a numerical expression to represent Phillip's total amount of money.

    $56 – $12

13. Ron's test scores are 75, 83, 86, and 79. What is his average for the four tests to the nearest whole number?

    81

14. Sally deposited $24.56 into her bank account. Let a represent the amount of money Sally had before the deposit. Write an algebraic expression for the money she has now.

    a + $24.56

15. There are 1 red, 1 green, 1 blue, and 1 yellow marker in a box. If you choose a marker without looking, what is the probability you will choose a blue marker?

    $\frac{1}{4}$

Name ____________________

Evaluating Numerical and Algebraic Expressions

Vocabulary

Complete.

1. To **evaluate** a numerical expression, perform the operations and write the expression as one number.

2. To evaluate an algebraic expression, replace the **variable** with a number and perform the operation in the expression.

Evaluate the numerical expression. Remember the order of operations.

3. $6 + 3 \times 8$ — **30**

4. $3.5 + 6.7 - 2.8$ — **7.4**

5. $9 \div 3 \times 4$ — **12**

6. $(3.5 - 1.7) \div 3$ — **0.6**

7. $5^2 + 3$ — **28**

8. $(15 + 9) \div 8$ — **3**

9. $\frac{2}{5} \times \frac{5}{8} + \frac{1}{4}$ — $\frac{1}{2}$

10. $3\frac{1}{2} \times (4 - 1\frac{1}{4})$ — $9\frac{5}{8}$

Evaluate the algebraic expression for the given value of the variable.

11. $a - 2$, for $a = 12$ — **10**

12. $c \div 7$, for $c = 42$ — **6**

13. $x + 19$, for $x = 28$ — **47**

14. $d \div (2 + 7)$, for $d = 54$ — **6**

15. $z^2 \times (17 - 5)$, for $z = 3$ — **108**

16. $56 \div (n + 2) - 5$, for $n = 5$ — **3**

Mixed Applications

17. The school is having a dance for the students. Regular admission is $5. Students with a VIP pass get $2 off the regular price. Write a numerical expression representing the cost of a ticket for a student with a VIP pass and evaluate it.

$5 − $2; $3

18. Jerry makes $0.25 for each magazine subscription he sells. If *m* represents the number of subscriptions he sold, what algebraic expression represents the amount of money he made? If he sold 110 subscriptions, how much did he make?

$0.25 × *m*; $27.50

19. There are 42 students in Tim's class. He wants to make 6 equal groups. How many students are in each group?

7 students

20. Selma bought $8.75 worth of treats to share with her friends. She paid for it with a $10 bill. What was her change?

$1.25

Use with text pages 280–281.

Name ______________________

LESSON 15.3

Input-Output Tables

Complete the input-output table.

1.

Input	Algebraic Expression	Output
x	$x + 3$	
7	7 + 3	10
8	8 + 3	11
9	9 + 3	12

2.

Input	Algebraic Expression	Output
a	$a - 8$	
13	13 − 8	5
15	15 − 8	7
19	19 − 8	11

Make an input-output table for the algebraic expression. Evaluate the expression for 2, 3, 4, and 5.

3. $b + 13$

b	$b + 13$	
2	2 + 13	15
3	3 + 13	16
4	4 + 13	17
5	5 + 13	18

4. $15 - g$

g	$15 - g$	
2	15 − 2	13
3	15 − 3	12
4	15 − 4	11
5	15 − 5	10

Determine the input for the given output. Make a table if necessary.

5. $x + 7$
output = 13
input = 6

6. $x - 11$
output = 20
input = 31

7. $x \div 6$
output = 9
input = 54

8. $a \times 8$
output = 24
input = 3

Mixed Applications

9. Timothy is used to getting 90% on his tests. The number he gets right is represented by the algebraic expression $90\% \times t$ or $0.9 \times t$, where t is the number of problems on his test. Make an input-output table to find out how many questions are on the test if 54 right answers equals 90%.

Check students' tables; 60 questions.

10. Lisa is saving for a bicycle that costs $184.75. She earns $65.00 a month mowing lawns. If m represents the number of months she works, what algebraic expression represents her earnings? Make an input-output table to find out how many months she must work to buy the bicycle.

$65.00 × m; check students' tables; 3 months.

11. What is the area of a rectangular garden with a length of 15 ft and a width of 10.5 ft?

157.5 ft^2

12. Stephie spent $45 at the mall. Emanuel spent $9 more than Stephie. Write a numerical expression to represent how much Emanuel spent.

$45 + $9

Name ______________________________

LESSON 15.4

Solving Addition and Subtraction Equations

Vocabulary

Complete.

1. Addition and subtraction are **inverse operations**.

Determine whether the given value is a solution of the equation. Write *yes* or *no*.

2. $x + 3 = 12; x = 5$ **no**

3. $17 + y = 23; y = 6$ **yes**

4. $t - 1 = 9; t = 10$ **yes**

5. $d - 1.5 = 7; d = 8.5$ **yes**

6. $13.4 + h = 17.8; h = 4.4$ **yes**

7. $4 = r - \frac{1}{3}; r = \frac{2}{3}$ **no**

Use inverse operations to solve. Check your solution.

8. $x + 6 = 8$ **$x = 2$**

9. $a + 9 = 13$ **$a = 4$**

10. $y - 4 = 19$ **$y = 23$**

11. $d - 20 = 35$ **$d = 55$**

12. $n + 7.3 = 20$ **$n = 12.7$**

13. $36.5 = t - 18.6$ **$t = 55.1$**

14. $24\frac{5}{8} = w + 15$ **$w = 9\frac{5}{8}$**

15. $f - 8\frac{3}{4} = 17$ **$f = 25\frac{3}{4}$**

Mixed Applications

For Exercises 16–17, choose a variable, write an equation, and solve it. **Variables may vary.**

16. Bill collected \$25.68 from his classmates for their end-of-the-year party. He gave \$15.68 to his teacher and used the rest to pay for expenses. How much did he use for expenses?
x = amount for expenses; $x + \$15.68 = \$25.68; x = \$10$

17. Tina used 35 balloons for the class party. There were 58 balloons left in the bag. How many balloons were there originally?
b = total number of balloons; $b - 35 = 58; b = 93$ balloons

18. The train leaves at 7:30 A.M. It takes Dana 20 min to walk to the train station and 15 min to eat breakfast there. When should Dana leave her house?
6:55 A.M.

19. Erik has \$45.00. He wants to buy a shirt for \$18.95, a hat for \$12.95, and a ball for \$7.50. Does he have enough money? Explain.
Yes; \$18.95 + \$12.95 + \$7.50 = \$39.40; \$39.40 < \$45.00

Use with text pages 286–289.

Name ______________________

LESSON 16.1

Solving Multiplication and Division Equations

Determine if the given value is a solution of the equation. Write *yes* or *no*.

1. $3y = 9; y = 3$ — yes

2. $4p = 24; p = 5$ — no

3. $\frac{x}{2} = 7; x = 14$ — yes

4. $\frac{s}{3} = 5; s = 12$ — no

5. $20 = 4n; n = 4$ — no

6. $32 = 8k; k = 4$ — yes

7. $7 = \frac{a}{9}; a = 63$ — yes

8. $4 = \frac{m}{8}; m = 2$ — no

Use inverse operations to solve. Check your solution.

9. $2x = 8$ — $x = 4$

10. $3c = 18$ — $c = 6$

11. $\frac{a}{4} = 8$ — $a = 32$

12. $\frac{m}{5} = 4$ — $m = 20$

13. $6 = \frac{k}{4}$ — $k = 24$

14. $60 = 5y$ — $y = 12$

15. $11 = \frac{b}{6}$ — $b = 66$

16. $45 = 3n$ — $n = 15$

17. $140 = 14g$ — $g = 10$

18. $513 = \frac{w}{3}$ — $w = 1,539$

19. $1,320 = 22d$ — $d = 60$

20. $19 = \frac{g}{11}$ — $g = 209$

Mixed Applications

For Exercises 21–23, choose a variable, write an equation, and solve. **Variables and equations will vary.**

21. Martin earns \$5 per hour. Last week he earned \$85. How many hours did he work last week?

h = hours worked; $5h = 85$; $h = 17$; 17 hr

22. Last night Maria spent $\frac{1}{4}$ of her homework time on mathematics. She spent 35 min on mathematics. How much time did she spend on homework?

m = minutes on homework; $\frac{m}{4} = 35$; $m = 140$;

140 min, or 2 hr 20 min

23. Renato gave his friend 5 candies. He has 31 candies left. How many candies did Renato have?

c = candies; $c - 5 = 31$; $c = 36$; 36 candies

24. Brenna and 3 friends share a pizza. If they share equally, what percent of the pizza does Brenna eat?

25%

Name ______________________________

LESSON 16.2

Money Relationships

Find the number of quarters, dimes, nickels, and pennies in the given dollar amount.

1. $6.00

   24; 60; 120; 600

2. $11.00

   44; 110; 220; 1,100

3. $23.00

   92; 230; 460; 2,300

4. $2.00

   8; 20; 40; 200

5. $7.50

   30; 75; 150; 750

6. $15.50

   62; 155; 310; 1,550

Use the formula $1.6261 \times P = D$ or $0.6200 \times M = D$ to convert the English pounds (P) or German marks (M) to U.S. dollars (D).

7. 20 pounds

   $32.52

8. 102 pounds

   $165.86

9. 50 marks

   $31.00

10. 225 marks

    $139.50

11. 19 pounds

    $30.90

12. 97 pounds

    $157.73

13. 28 marks

    $17.36

14. 187 marks

    $115.94

Mixed Applications

15. Jean emptied her piggy bank. She counted $19.75 in pennies, nickels, dimes, and quarters. There were 25 pennies, 40 nickels, and 60 quarters. How many dimes were there?

    25 dimes

16. Hank estimates that a tree grows about 2 ft per year. The tree is 9 ft tall now. How tall will it be in 4 yr?

    17 ft tall

17. Izeta bought some groceries and received all her change in quarters. She received $1.75 in change. How many quarters did she receive?

    7 quarters

18. Mike practiced piano for 45 min. This was $\frac{1}{3}$ the time he spent doing homework. How much time did Mike spend doing homework? Choose a variable, write an equation, and solve.

    Variables and equations may vary. h = time doing homework; $\frac{h}{3} = 45$; $h = 135$; 135 min, or 2 hr 15 min

Use with text pages 300–301.

Name ______________________

LESSON 16.3

Temperature Relationships

Convert the temperature from degrees Celsius to degrees Fahrenheit. Round the answer to the nearest degree.

1. 30°C	2. 25°C	3. 50°C	4. 13°C	5. 3°C	6. 60°C
86°F	77°F	122°F	55°F	37°F	140°F
7. 22°C	8. 54°C	9. 7°C	10. 100°C	11. 15°C	12. 0°C
72°F	129°F	45°F	212°F	59°F	32°F
13. 38°C	14. 59°C	15. 14°C	16. 5°C	17. 8°C	18. 61°C
100°F	138°F	57°F	41°F	46°F	142°F

Convert the temperature from degrees Fahrenheit to degrees Celsius. Round the answer to the nearest degree.

19. 71°F	20. 50°F	21. 140°F	22. 90°F	23. 45°F	24. 121°F
22°C	10°C	60°C	32°C	70°C	49°C
25. 32°F	26. 49°F	27. 96°F	28. 130°F	29. 113°F	30. 86°F
0°C	9°C	36°C	54°C	45°C	30°C
31. 40°F	32. 82°F	33. 73°F	34. 69°F	35. 43°F	36. 80°F
4°C	28°C	23°C	21°C	6°C	27°C

Mixed Applications

37. Jack lives near San Francisco. The temperature at his home was 88°F, but at the beach it was only 66°F. What was the temperature at the beach in degrees Celsius?

19°C

38. The temperature fell from 12°C to 1°C one evening in Chicago. By how many degrees Fahrenheit did the temperature drop?

20°F

39. Mike saves dimes. He has $7.80 in dimes. How many dimes does Mike have?

78 dimes

40. How many square numbers are there between 1 and 500?

22 square numbers;
$22^2 = 484$; $23^2 = 529$

Name ______________________________

Time and Distance Relationships

Use the formula $d = r \times t$ to complete.

1. d = 80 mi
r = 20 mi per hr
t = 4 hr

2. d = 714 ft
r = 17 ft per sec
t = 42 sec

3. d = 51.94 km
r = 9.8 km per hr
t = 5.3 hr

4. d = 75 mi
r = 25 mi per hr
t = 3 hr

5. d = 1,320 km
r = 6 km per min
t = 220 min

6. d = 99 ft
r = 9 ft per sec
t = 11 sec

7. d = 605 mi
r = 55 mi per hr
t = 11 hr

8. d = 336 ft
r = 28 ft per sec
t = 12 sec

9. d = 500 ft
r = 25 ft per min
t = 20 min

10. d = 351 mi
r = 39 mi per hr
t = 9 hr

11. d = 96 km
r = 8 km per min
t = 12 min

12. d = 2,025 ft
r = 15 ft per sec
t = 135 sec

13. d = 11,160,000 mi
r = 186,000 mi per sec
t = 60 sec

14. d = 95,000 ft
r = 95,000 ft per sec
t = 1 sec

15. d = 98.8 km
r = 26 km per hr
t = 3.8 hr

Mixed Applications

16. A spider walks across the floor at a constant rate of 2 ft per min. How long does it take the spider to walk 9 ft?

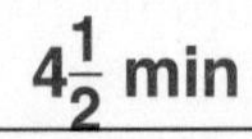

$4\frac{1}{2}$ min

17. Carlos has 25 coins whose total value is $1.00. What are the coins?

1 quarter, 3 dimes, 6 nickels, 15 pennies

18. If Angela walks at a constant, brisk pace, she can hike 10 mi in 3 hr. At what rate does she hike? Round the answer to the nearest whole number.

3 mi per hr

19. Cynthia keeps the temperature inside her apartment at 68°F during the winter. What is the temperature in degrees Celsius?

20°C

Use with text pages 304–305.

Name ______________________________

LESSON 16.5

Problem-Solving Strategy

Making a Table to Relate Measurements

Solve by making a table. **Check students' tables.**

1. Lailing bought 14 in. of beading on Tuesday and 60 cm on Wednesday. How much more beading did she buy on Wednesday than on Tuesday? (1 in. = 2.54 cm)

   24.44 cm more

2. Sarah is making modeling dough. The recipe calls for 3 tbsp of salt for every $\frac{1}{2}$ c of flour. How much salt does Sarah need for 3 c of flour?

   18 tbsp

Mixed Applications

Solve.

CHOOSE A STRATEGY

• Make a Table • Use Estimation • Write an Equation • Guess and Check • Find a Pattern • Work Backward

Choices of strategies will vary.

3. Rebecca baby-sat 25 hr over a 4-wk period. She baby-sat 6 hr the fourth week, 3 hr the third week, and 8 hr the second week. How many hours did she baby-sit the first week?

   8 hr

4. Matthew sews patches on his team's uniforms. He can sew 7 patches in 1 hr. How long will it take him to sew 175 patches?

   25 hr

5. A pattern has 2 flowers in the first row, 5 in the second row, 9 in the third row, and 14 in the fourth row. How many flowers are in the seventh row?

   35 flowers

6. The perimeter of a rectangle is 30 m. If the length is twice the width, what are the dimensions?

   5 m by 10 m

7. Tom can hike an average of 12 mi per day. Can he hike from Branchville to Westerville in 4 days?

   no

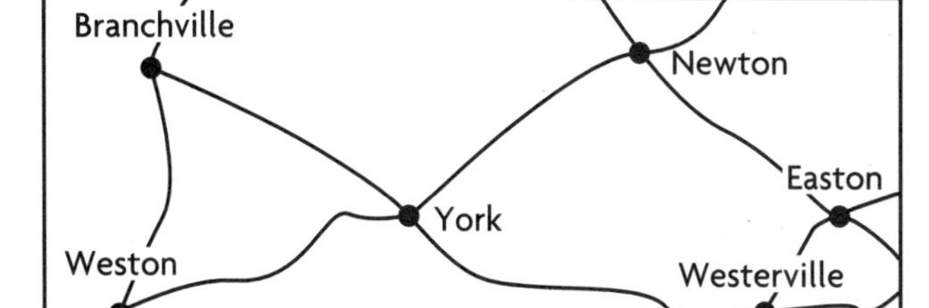

Name ______________________

Ratios

Vocabulary

Complete.

1. A **ratio** is a comparison of two numbers.

2. **Equivalent ratios** are ratios that name the same comparison.

Write the ratio in three ways.

3. seven to nine **7 to 9, 7:9, $\frac{7}{9}$**

4. twelve to one **12 to 1, 12:1, $\frac{12}{1}$**

5. seven to three **7 to 3, 7:3, $\frac{7}{3}$**

6. eight to six **8 to 6, 8:6, $\frac{8}{6}$**

7. five to two **5 to 2, 5:2, $\frac{5}{2}$**

8. eight to seventeen **8 to 17, 8:17, $\frac{8}{17}$**

For Exercises 9–10, use the figure at the right.

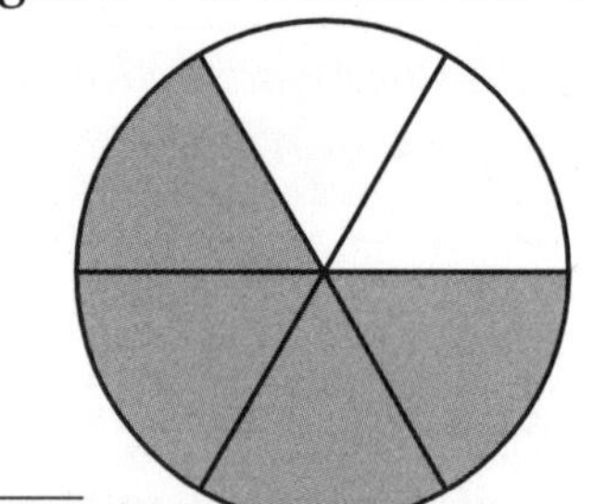

9. Find the ratio of shaded sections to unshaded sections. Then write three equivalent ratios.

   4 to 2; possible answers: 2:1, 8:4, 12:6

10. Find the ratio of shaded sections to all the sections. Then write three equivalent ratios.

   4 to 6; possible answers: 2:3, 8:12, 12:18

Find the missing term that makes the ratios equivalent.

11. $\frac{3}{7}, \frac{\square}{14}$ **6**

12. 7 to 5, $\square$ to 15 **21**

13. 15:5, 3:$\square$ **1**

Mixed Applications

14. A punch recipe calls for 3 c of orange juice and 2 c of grapefruit juice. What is the ratio of orange juice to grapefruit juice? to the mixed juice?

   3:2, 3:5

15. Tina spends \$0.95 for a glass of milk and \$0.85 for a muffin. She pays with a \$5 bill. What is her change?

   \$3.20

Use with text pages 316–317.

Name ______________________________

Rates

Vocabulary

Complete.

1. A **rate** is a ratio that compares two quantities having different units of measure.
2. When the second term of a rate is 1, the rate is called a **unit** rate.

Write a ratio in fraction form for each rate.

3. 7 apples for $1.00 — $\frac{\text{7 apples}}{\text{\$1.00}}$
4. $0.06 per page — $\frac{\text{\$0.06}}{\text{1 page}}$
5. 24 people in 6 cars — $\frac{\text{24 people}}{\text{6 cars}}$
6. 65 mi per 3 gal — $\frac{\text{65 mi}}{\text{3 gal}}$
7. 5 CDs for $49 — $\frac{\text{5 CDs}}{\text{\$49}}$
8. $20 per dozen tarts — $\frac{\text{\$20}}{\text{12 tarts}}$

Write the unit rate in fraction form.

9. $8.00 for 10 — $\frac{\text{\$0.80}}{1}$
10. $2.40 per dozen — $\frac{\text{\$0.20}}{1}$
11. 100 students in 4 classrooms — $\frac{\text{25 students}}{\text{1 classroom}}$
12. 220 mi per 11 gal — $\frac{\text{20 mi}}{\text{1 gal}}$
13. $5.00 for 25 — $\frac{\text{\$0.20}}{1}$
14. $280 for 7 — $\frac{\text{\$40}}{1}$

Mixed Applications

15. A box of cereal weighs 550 g. There is enough cereal for 10 servings. How many grams of cereal are there per serving?

    55 g

16. One store has oranges on sale, 8 for $1.60. Another store sells oranges for $2.60 per dozen. Which store has the better buy? Explain.

    8 for $1.60; $0.20 per orange < about $0.22 per orange

17. The total surface area to be painted is 1,750 square feet. A gallon of paint covers 400 square feet. About how many gallons of paint are needed?

    about 5 gal

18. Telly buys 3 shirts for $12.95 each, a tie for $9.50, and 2 belts at $17.75 each. How much does he spend?

    $83.85

Name ______________________

LESSON 17.3

Percents

Vocabulary

Complete.

1. A ___percent___ is the ratio of a number to 100.

Write the ratio as a percent.

2. $\frac{65}{100}$ ___65%___ 3. 22:100 ___22%___ 4. $\frac{80}{100}$ ___80%___ 5. 17:100 ___17%___

Write the decimal as a percent.

6. 0.7 ___70%___ 7. 0.18 ___18%___ 8. 0.84 ___84%___ 9. 0.41 ___41%___

Write the ratio as a percent.

10. $\frac{3}{5}$ ___60%___ 11. $\frac{17}{100}$ ___17%___ 12. $\frac{5}{8}$ ___62.5%___ 13. $\frac{8}{25}$ ___32%___

Write the percent as a decimal and as a ratio in simplest form.

14. 75% ___0.75; $\frac{3}{4}$___ 15. 30% ___0.3; $\frac{3}{10}$___ 16. 55% ___0.55; $\frac{11}{20}$___

17. 56% ___0.56; $\frac{14}{25}$___ 18. 84% ___0.84; $\frac{21}{25}$___ 19. 6% ___0.06; $\frac{3}{50}$___

Complete by using $<$, $>$, or $=$. If $<$ or $>$, tell why.

20. $\frac{1}{10}$ $\boxed{=}$ 10% 21. 65% $\boxed{>}$ $\frac{30}{50}$ 22. 2.5% $\boxed{=}$ 0.025

______________ $\frac{30}{50} = 60\%$ ______________

Mixed Applications

23. Ellie answered 17 out of 20 questions correctly on her last math test. What percent of the questions did she answer correctly?

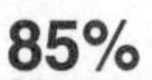

___85%___

24. The test started at 9:44 A.M. and ended at 11:14 A.M. How many hours did the test last?

___$1\frac{1}{2}$ hr___

25. A used car costs $4,000. The tax is $200. What percent is the tax?

___5%___

26. A dozen roses cost $60.00. What is the cost of two roses?

___$10.00___

Use with text pages 320–323.

Name ______________________

LESSON 17.4

Percents and the Whole

Tell what percent of the figure is shaded.

1. 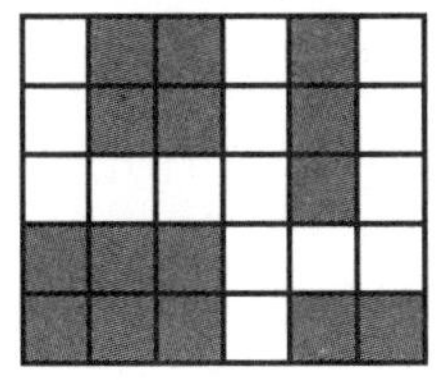

50%

2. 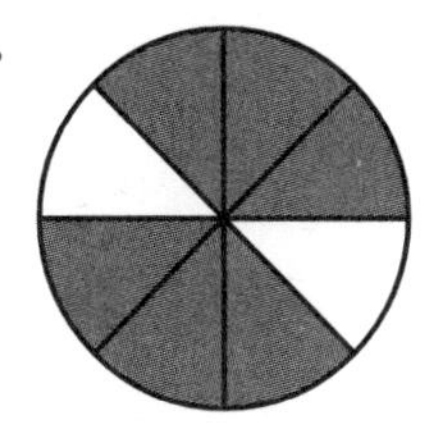

75%

3. 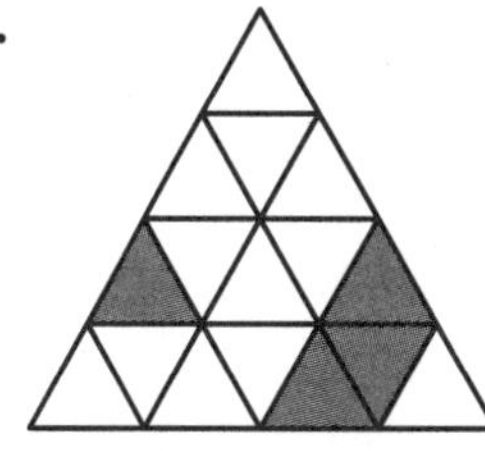

25%

4. 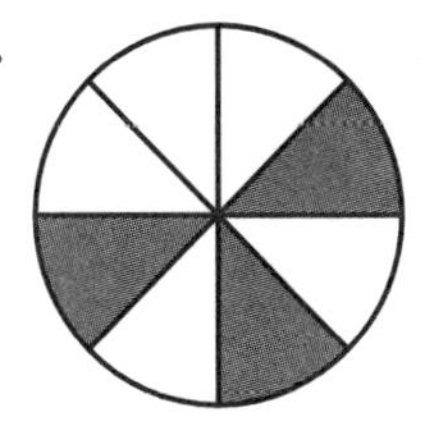

37.5%

5. 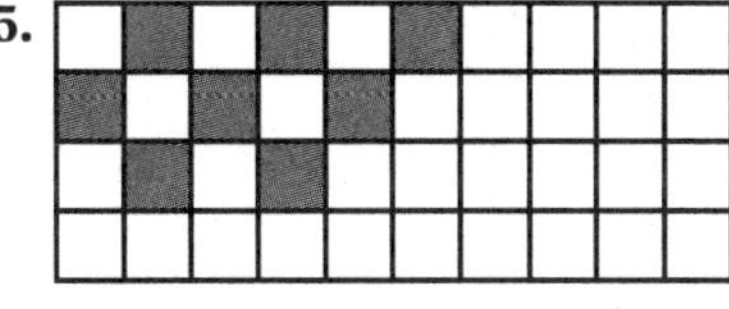

20%

6. 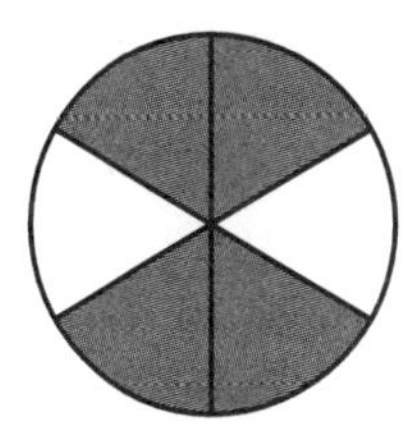

$66\frac{2}{3}\%$

7. 

87.5%

8. 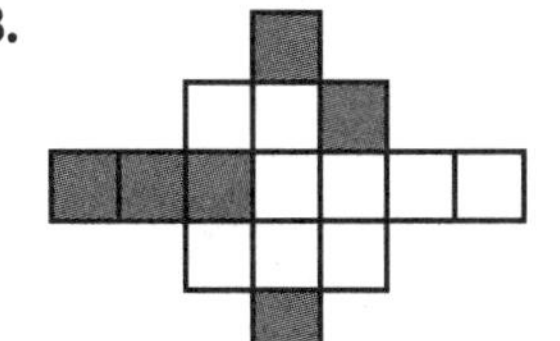

40%

9. 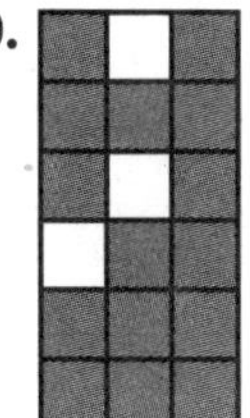

$83\frac{1}{3}\%$

Mixed Applications

10. The game of checkers is played on a board with 64 squares. At the start of the game, 24 squares are occupied. What percent of the squares are occupied?

37.5%

11. A window has six rows of panes, with 4 panes in each row. Three of the panes are cracked. What percent of the panes are cracked?

12.5%

12. Martin's Clothes is having a half-price sale. A suit normally sells for $86.50. What is the sale price?

$43.25

13. A garden is next to a garage. The garden is fenced in, with the garage forming one of the four sides. The garden is 24 ft by 20 ft. How many feet of fencing do you need?

64 ft or 68 ft, depending on length of garage

Name ______________________________

LESSON 17.5

Problem-Solving Strategy

Write a Proportion

Write a proportion to solve.

1. A dripping faucet wastes 3 cups of water in 24 hr. How much water is wasted in 56 hours?

   7 c

2. A map uses the scale of 3 cm for every 10 km. If the map shows a distance of 12 cm, what is the actual distance?

   40 km

3. A pump empties the pool at the rate of 1,000 gal every 4 hours. How long does it take to pump out 20,000 gallons?

   80 hr

4. Tom drinks 8 oz of water for every 3 mi he bikes. After 21 miles, how much water has he drunk?

   56 oz

5. A punch consists of 2 parts ginger ale and 3 parts orange juice. If the punch bowl contains 8 c of ginger ale, how many cups of punch are in the bowl?

   20 c

6. A 5-lb bag of apples contains 12 apples. What will a bag of 40 apples weigh?

   $16\frac{2}{3}$ lb

Solve.

CHOOSE A STRATEGY

- Find a Pattern
- Write a Proportion
- Write an Equation
- Work Backward
- Use a Formula
- Guess and Check

Choices of strategies will vary.

7. Sol has 22 coins, all nickels and dimes. The coins are worth a total of $1.55. How many nickels are there?

   13 nickels

8. Two sides of a triangle are the same length. One side measures $3\frac{1}{2}$ ft; another measures 2 ft. Give two possible perimeters of the triangle.

   $7\frac{1}{2}$ ft or 9 ft

9. The volume, V, of a prism is found using the formula $V = B \times h$, where B is the area of the base, and h is the height of the prism. Find the volume of a prism with a 20-cm height, if the area of its base is 200 cm^2.

   4,000 cm^3

10. A 2-in. × 2-in. square is entirely contained within a 6-in. × 6-in. square. You shade the area that is within the larger square but not in the smaller square. What is the area of the shaded part?

    32 in.^{2}

Use with text pages 328–329.

Name ______________________________

LESSON 18.1

Problem-Solving Strategy

Acting It Out to Find a Percent of a Number

Solve the problem by acting it out.

1. The Children's Toy Store sold 500 games in June. In July it sold 40% of that amount. How many games did it sell in July?

   200 games

2. The National Pizza Company has sold 10 million pizzas during the last year. Of the pizzas sold, 70% of those were with pepperoni only. How many pizzas did it sell with pepperoni only?

   7 million pizzas

3. A game-show host has done 2,000 shows. He did 30% of those in Los Angeles. How many shows did he do in Los Angeles?

   600 shows

4. Mr. Hammer paid an interest rate of 20% last year on his $20,000 car. How much money did he pay in interest for the year?

   $4,000

Mixed Applications

Solve.

CHOOSE A STRATEGY

- Draw a Diagram • Act It Out • Work Backward • Write an Equation • Guess and Check • Make a Table

Choices of strategies will vary.

5. Mr. Kincaid gets a 10% bonus for every $500,000 worth of cars he sells. How much is his bonus for $500,000 worth of cars?

   $50,000

6. A local musical group has played for 5 years. During that time they have performed 1,585 times. On average, how many times have they performed each year?

   317 times

7. Tickets for a local sporting event cost $6 for children and $15 for adults. Ms. Jacobson bought 23 tickets for $228. How many of each kind of ticket did she buy?

   10 adult and 13 child tickets

8. Sylvia is walking home 6 miles from her friend's house. She walks at 3 mph. How long will it take her to get home if she stops at the park for an hour on the way home?

   3 hours

9. Bill went to the florist. He bought a bouquet of roses for $24 and two bouquets of carnations for $8 each. He has $15 left. How much money did he take to the florist?

   $55

10. The Davidson family uses 2,000 gal of water each month. They plan on saving 30% next month. How many gallons of water do they plan on saving?

    600 gal

Name ______________________

LESSON 18.2

Percent of a Number

Use a ratio in simplest form to find the percent of the number.

1. 10% of 8	2. 25% of 60	3. 50% of 50	4. 70% of 90	5. 80% of 70
$\frac{4}{5}$	15	25	63	56

Use a decimal to find the percent of the number.

6. 15% of 8	7. 35% of 45	8. 55% of 92	9. 82% of 70	10. 93% of 24
1.2	15.75	50.6	57.4	22.32

Use the method of your choice to find the percent of the number.

11. 18% of 15	12. 22% of 60	13. 25% of 64	14. 30% of 75	15. 4% of 30
2.7	13.2	16	22.5	1.2
16. 52% of 40	17. 96% of 84	18. 81% of 34	19. 12% of 300	20. 67% of 200
20.8	80.64	27.54	36	134
21. 4.5% of 90	22. 110% of 30	23. 140% of 100	24. 200% of 250	25. 400% of 80
4.05	33	140	500	320

Find the sales tax. Round to the nearest cent when necessary.

26. price: $30 tax rate: 8%	27. price: $15.80 tax rate: 11%	28. price: $6.50 tax rate: 7.5%	29. price: $120.00 tax rate: 4%
$2.40	$1.74	$0.49	$4.80

Mixed Applications

30. Helen wants to buy an alarm clock that costs $28. She has $35. The sales tax rate is 9%. Does she have enough money for the alarm clock and sales tax? How much is the sales tax?

Yes; the tax on her purchase is $2.52.

31. Matt bought a poster for $8.95, a cap for $13.50, and a notebook for $3.95. If he gives the cashier $30, how much change will he get?

$3.60

32. Derrick bought a gift on layaway. His gift cost $85. The store required a 30% deposit to hold the item. How much is the deposit?

$25.50

33. Debra is carpeting her room. The room is $12\frac{3}{4}$ ft long and $10\frac{1}{2}$ ft wide. How many square feet of carpeting does she need?

$133\frac{7}{8}$ ft^2

Use with text pages 336–339.

Name ______________________________

Circle Graphs

For Exercises 1–2, use the data in the table at the right.

1. Find the angles you would use to make a circle graph.

10% = 36°, 50% = 180°,

25% = 90°, 15% = 54°

BUDGET	
Item	**Percent**
Gas	10%
Clothes	50%
Food	25%
Other	15%

2. Make a circle graph.

BUDGET

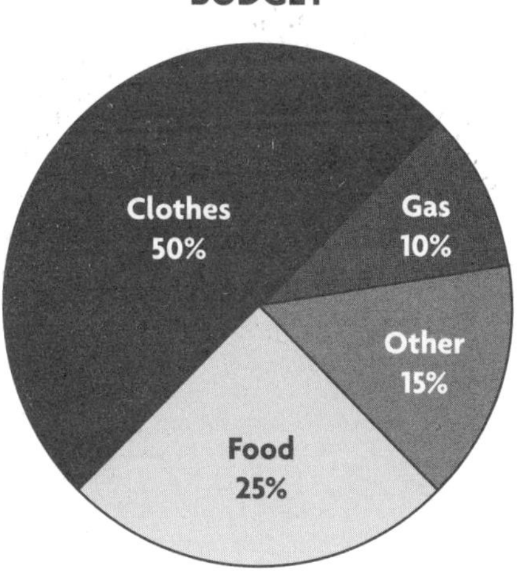

Problem 5

MOST POPULAR CAR MODELS

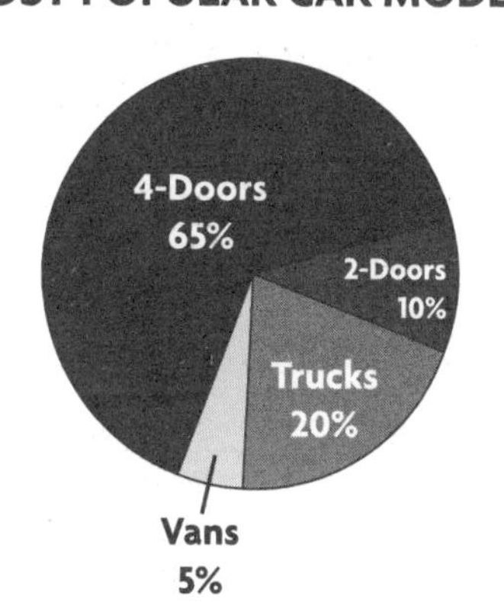

Problem 6

SALINA'S MONEY

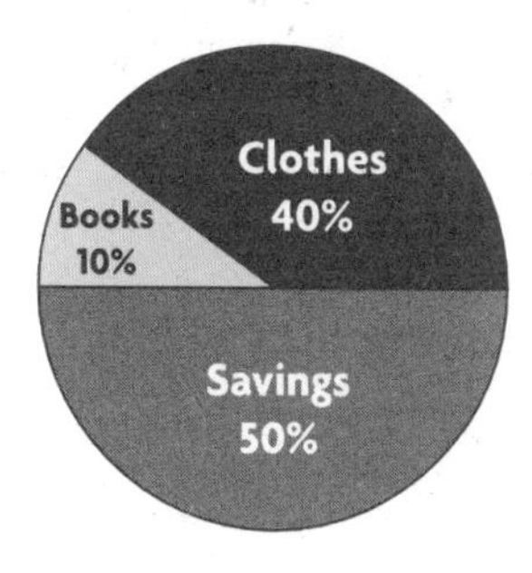

For Exercises 3–4, use the circle graph at the right.

3. If 400 people were surveyed, how many like rock?

168 people

4. If 800 people were surveyed, how many like jazz the most?

224 people

MUSIC PEOPLE LISTEN TO MOST

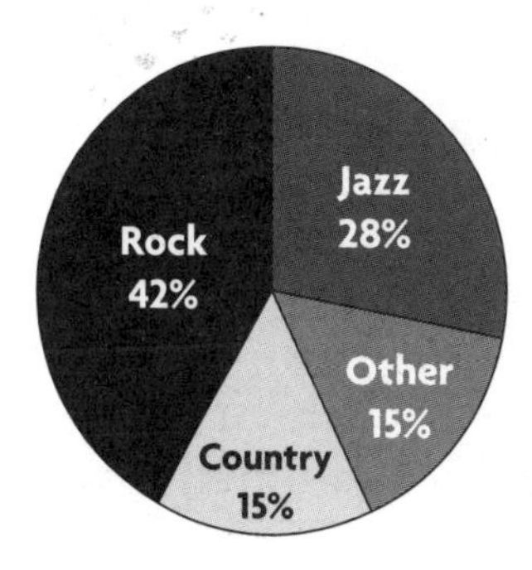

Mixed Applications

5. In a survey, people were asked about their favorite type of car. Of 100 people, 65% chose 4-doors, 5% chose vans, 20% chose trucks, and 10% chose 2-doors. Use the results to make a graph. **See graph above.**

6. Salina had \$300. She spent 40% on clothes and 10% on books. She put the rest in her savings account. Make a circle graph to show how Salina spent her \$300. How much did she put in her savings account? **See graph above.**

\$150

Name ____________________

LESSON 18.4

Discount

Vocabulary

Complete.

1. To find the amount of the ____discount____, multiply the regular price by the discount rate.

Find the amount of discount.

2. regular price: $68.00 — 30% off — **$20.40**

3. regular price: $50.00 — SAVE 60% — **$30.00**

4. regular price: $54.00 — 20% Discount — **$10.80**

5. regular price: $26.00 — 5% off — **$1.30**

Find the sale price.

6. regular price: $18.50 — Discount 20% — **$14.80**

7. regular price: $35.00 — 25% off — **$26.25**

8. regular price: $45.50 — SAVE 50% — **$22.75**

9. regular price: $23.60 — SALE 80% off — **$4.72**

Mixed Applications

10. Wesley bought a book for $45. The next day, he noticed that it was on sale for 40% off. He took the book back to the store and got a refund of 40%. How much was his refund?

    $18

11. Doug has two pieces of wood. One piece is $\frac{4}{6}$ yd long, and the other is $\frac{2}{3}$ yd long. How many yards of wood does he have altogether?

    $1\frac{1}{3}$ yd

12. Lisa has saved $28. She is waiting for a jacket to go on sale for 20% off its regular price of $50. How much more does she have to save to buy the jacket at the sale price?

    $12

13. Cheryl is having a party for 68 people. She already has 4 tables which seat 5 people. A party-supply company rents tables which seat 6. How many tables will she need to rent?

    8 tables

Use with text pages 342–343.

Name ______________________

LESSON 18.5

Simple Interest

Vocabulary

Complete.

1. Simple interest is the amount of interest earned on the amount deposited.
2. The amount deposited is called the principal.

Find the interest. Round to the nearest cent where necessary.

	Principal	Yearly Rate	Interest for 1 Year	Interest for 2 Years
3.	$80	3%	$2.40	$4.80
4.	$150	4.5%	$6.75	$13.50
5.	$340	6%	$20.40	$40.80
6.	$600	5.2%	$31.20	$62.40
7.	$1,400	7.9%	$110.60	$221.20
8.	$5,500	9%	$495.00	$990.00
9.	$7,500	8.5%	$637.50	$1,275.00
10.	$10,000	9.6%	$960.00	$1,920.00
11.	$11,350	9.8%	$1,112.30	$2,224.60
12.	$12,975	9.5%	$1,232.63	$2,465.25

Mixed Applications

13. Jill put $1,200 in the bank for 1 year. She earned interest at the rate of 5% a year. How much did she earn?

$60

14. Harry had $200 that earned 3.5% interest a year in a bank. After 2 years, he withdrew his money. How much did he withdraw?

$214

15. George is waiting for a bike he wants to go on sale. He was told that it will go on sale for 35% off the regular price of $580. How much will he save if he waits for the sale? What will the sale price be?

$203; $377

16. Brenda weeded $\frac{1}{5}$ of her garden on Monday, $\frac{1}{2}$ on Tuesday, and $\frac{3}{10}$ on Wednesday. Did she weed the entire garden? Explain.

yes; $\frac{1}{5} + \frac{1}{2} + \frac{3}{10} = 1$

Name ______________________________

LESSON 19.1

Similar and Congruent Figures

Look at each figure. Tell whether each pair of shapes appear to be *similar, congruent, both,* or *neither.*

1. 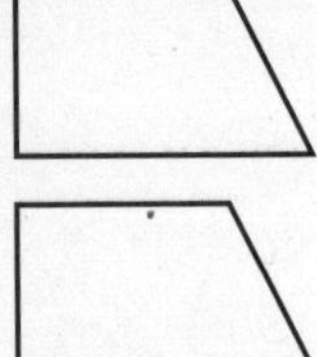

both

2. 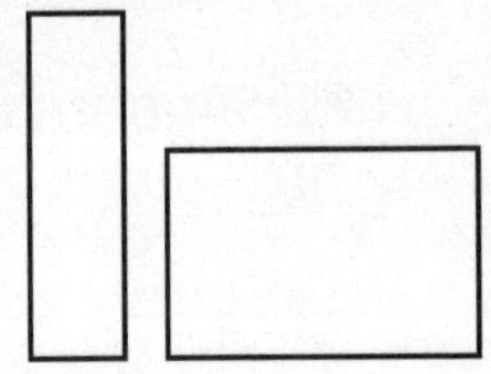

neither

3. 

similar

4. 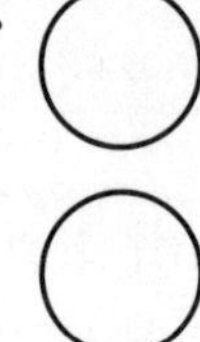

both

5. 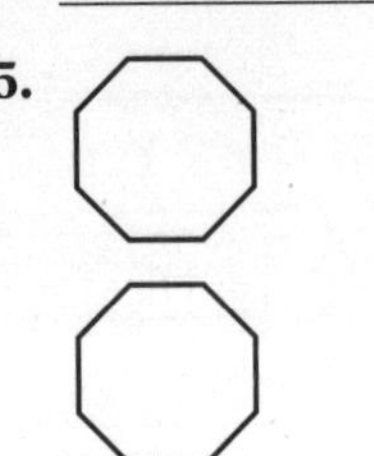

both

6. 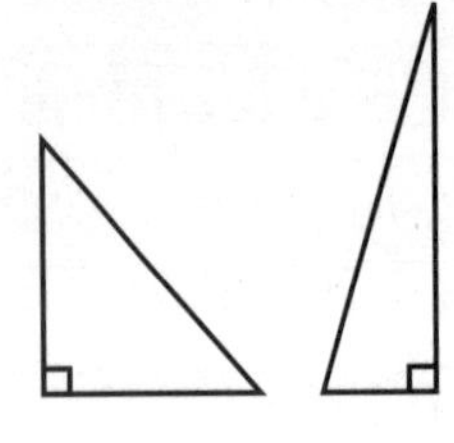

neither

7. 

similar

8. 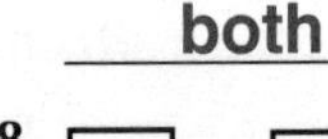
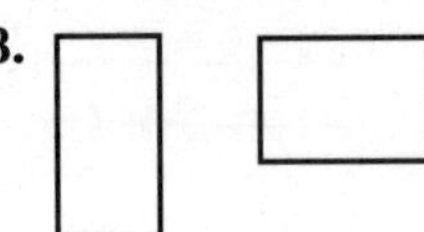

neither

Mixed Applications

Use the diagram to answer Exercises 9–12.

9. How many pairs of similar shaded squares can you find?

10

10. How many pairs of congruent shaded squares can you find?

6

11. How many pairs of similar blank squares can you find?

10

12. How many pairs of congruent blank squares can you find?

6

13. How many $1\frac{1}{4}$-ft pieces of wood can be cut from a $12\frac{1}{2}$-ft piece of wood?

10 pieces

14. If you have a cube numbered 1 through 6, what is the probability you will roll a number less than 4?

$\frac{1}{2}$

Use with text pages 354–355.

Name ____________________

LESSON 19.2

Ratios and Similar Figures

Vocabulary

Complete.

1. Similar figures have corresponding **sides** and corresponding **angles**.

Name the corresponding sides and angles. Write the ratio of the corresponding sides in simplest form.

2. 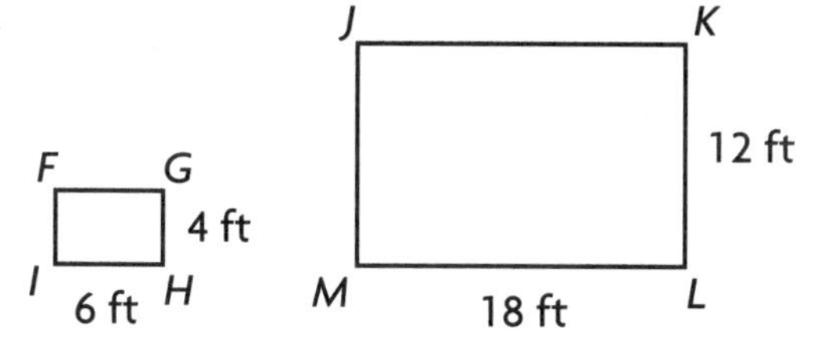

$\overline{FG}$ corresponds to $\overline{JK}$; $\overline{GH}$ corresponds to $\overline{KL}$; $\overline{HI}$ corresponds to $\overline{LM}$; $\overline{IF}$ corresponds to $\overline{MJ}$; $\angle F$ corresponds to $\angle J$; $\angle G$ corresponds to $\angle K$; $\angle H$ corresponds to $\angle L$; $\angle I$ corresponds to $\angle M$; $\frac{1}{3}$, or $\frac{3}{1}$

3. 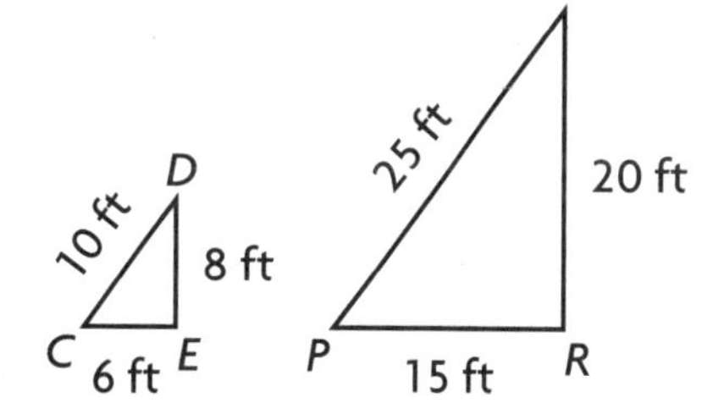

$\overline{CD}$ corresponds to $\overline{PQ}$; $\overline{DE}$ corresponds to $\overline{QR}$; $\overline{EC}$ corresponds to $\overline{RP}$; $\angle C$ corresponds to $\angle P$; $\angle D$ corresponds to $\angle Q$; $\angle E$ corresponds to $\angle R$; $\frac{2}{5}$, or $\frac{5}{2}$

Tell whether the figures in each pair are similar. Write *yes* or *no*. If you write *no*, explain.

4. 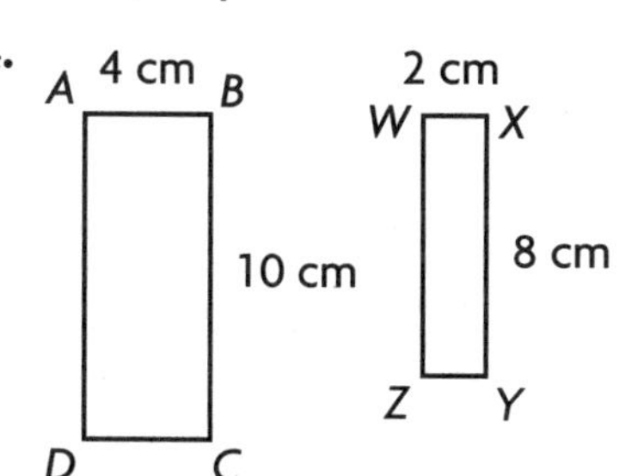

No; ratios are not equivalent.

5. 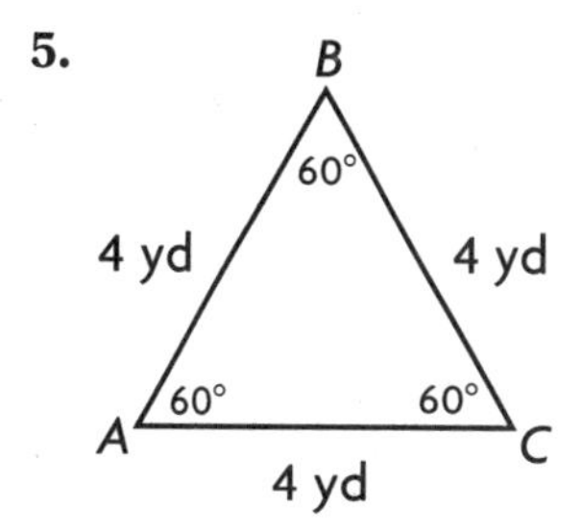

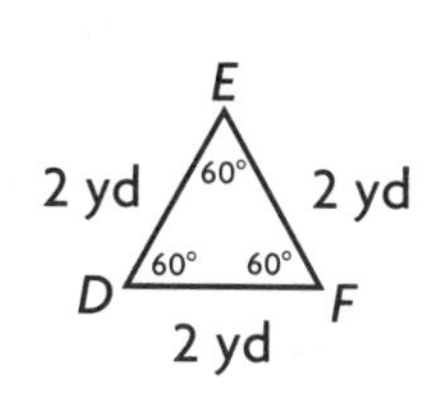

yes

Mixed Applications

6. Tony bought two different sizes of index cards for school. They were 3 in. by 5 in. and 5 in. by 7 in. Are the two types of index cards similar?

no

7. Steve has $125. He puts $85 in a savings account and the rest in a checking account. How much does he put in the checking account?

$40

Use with text pages 356–359.

Name ______________________________

LESSON 19.3

Proportions and Similar Figures

The figures in each pair are similar. Find n.

1.

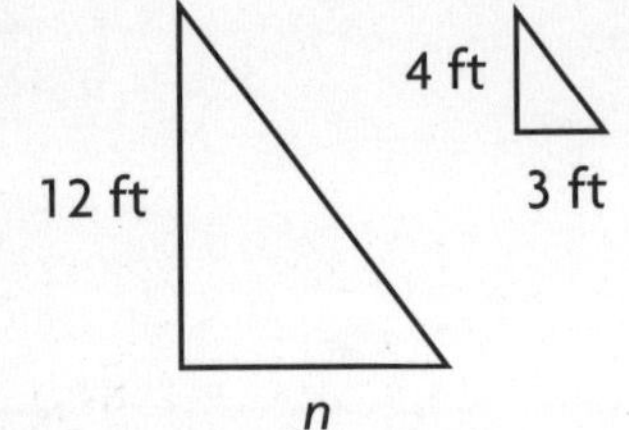

$n = 9$ ft

2.

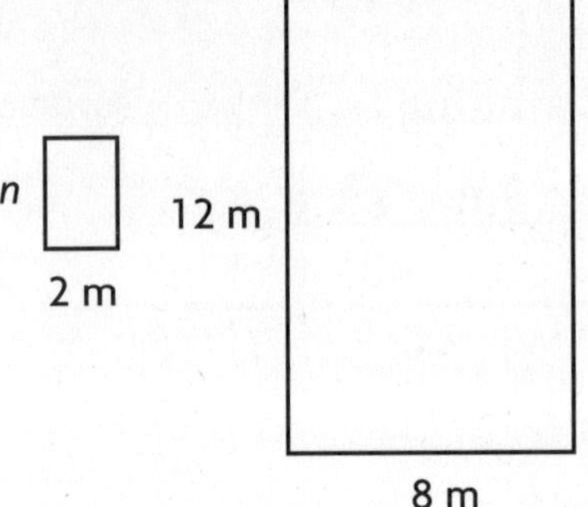

$n = 3$ m

3.

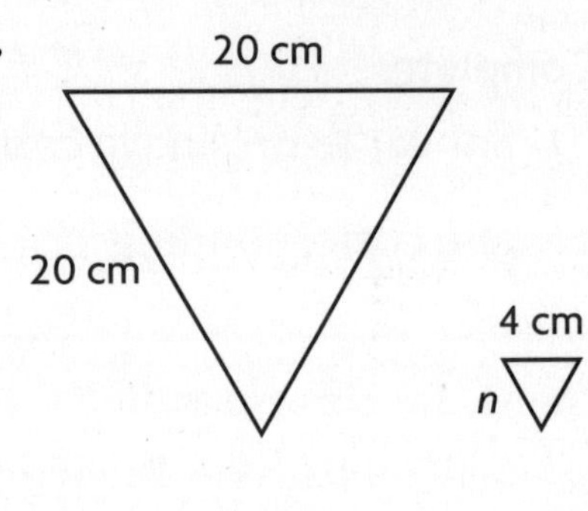

$n = 4$ cm

4.

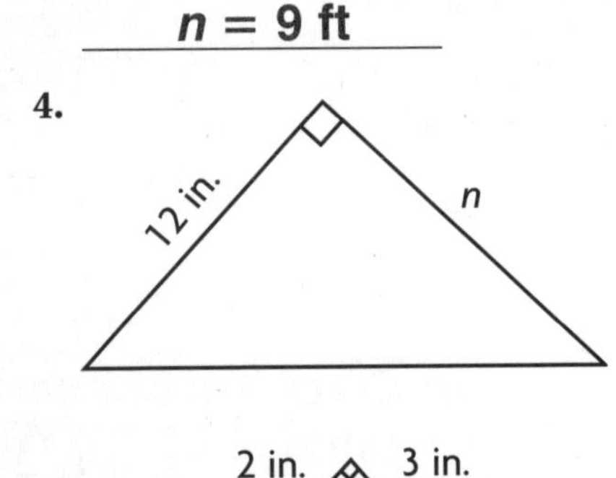

$n = 18$ in.

5.

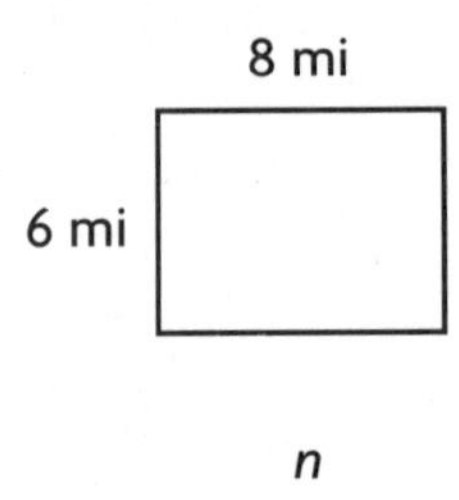

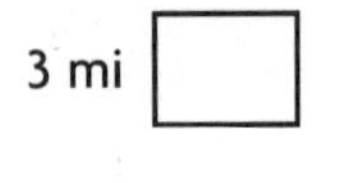

$n = 4$ mi

6.

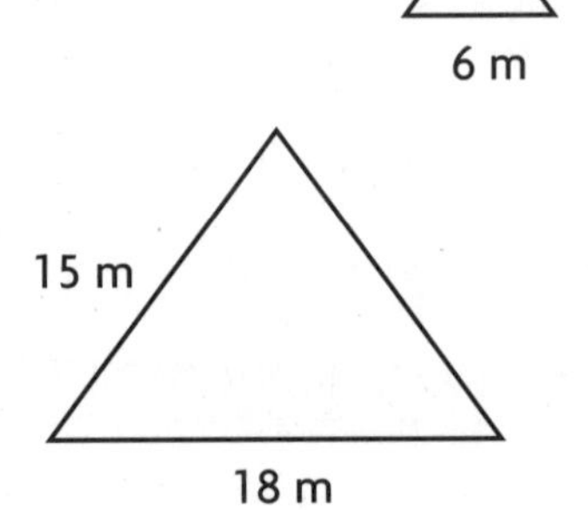

$n = 5$ m

Mixed Applications

7. Tanya has a newspaper photo that she wants to enlarge to include in her report. The photo is 3 in. by 4 in. She wants to enlarge it so that it is 8 in. long. How wide should she make the photo?

   6 in.

8. Sally bought some clothes on lay-away. The total was \$64. The store required a 25% deposit to hold the item. How much is the deposit?

   \$16

9. Jack is making a scale replica of an old-style home. The house is 30 ft long. The width of the house is 20 ft. He wants his model to be 2 ft wide. How long should he make his model?

   3 ft

10. Kirk is recycling cans to raise the money for his vacation. He gets \$0.15 for each pound. Write an algebraic expression for how much he makes for each pound, p, he collects. How much would he make if he collected 240 pounds?

    $0.15p$; \$36.00

Use with text pages 360–361.

Name ______________________________

LESSON 19.4

Proportions and Indirect Measurement

Vocabulary

Complete.

1. Using similar figures and proportions to find a measurement is called using **indirect measurement**.

Find the unknown length.

2.

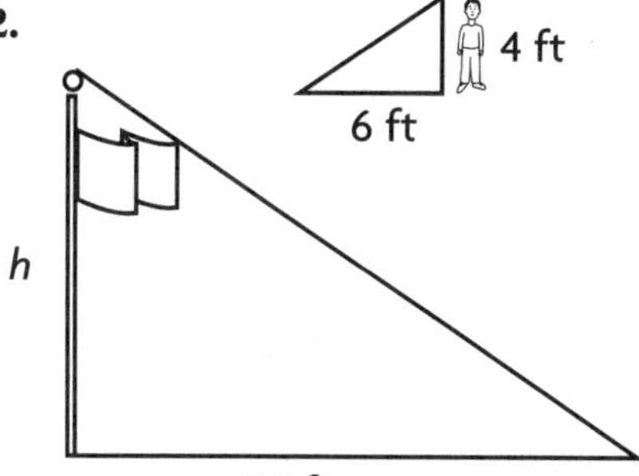

$\frac{4}{h} = \frac{6}{24}$; $h = 16$ ft

3.

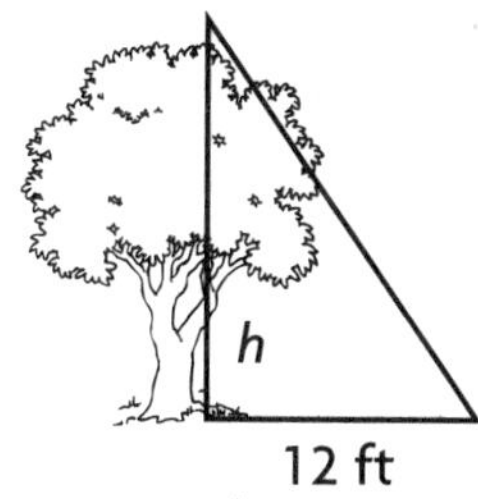

$\frac{3}{h} = \frac{2}{12}$; $h = 18$ ft

4.

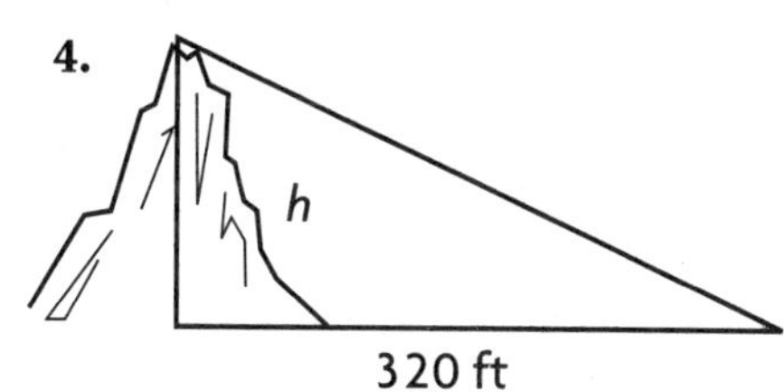

$\frac{5}{h} = \frac{10}{320}$; $h = 160$ ft

5.

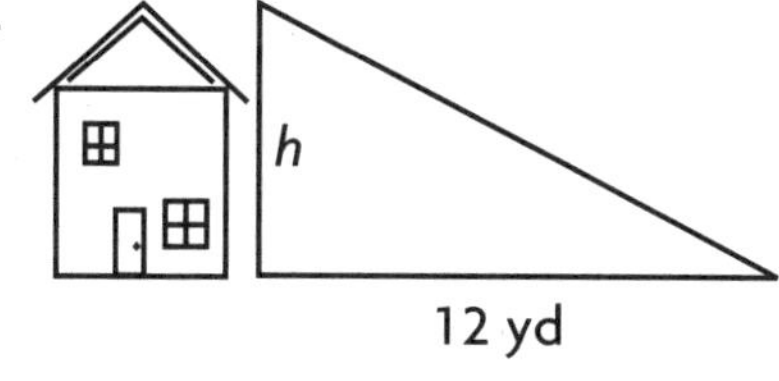

$\frac{1}{h} = \frac{2}{12}$; $h = 6$ yd

6.

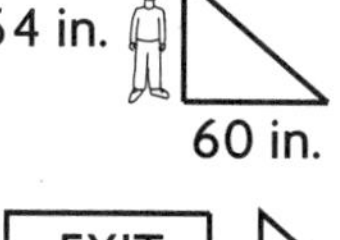

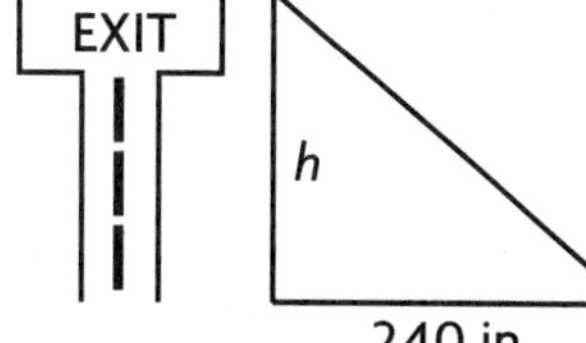

$\frac{54}{h} = \frac{60}{240}$; $h = 216$ in.

7.

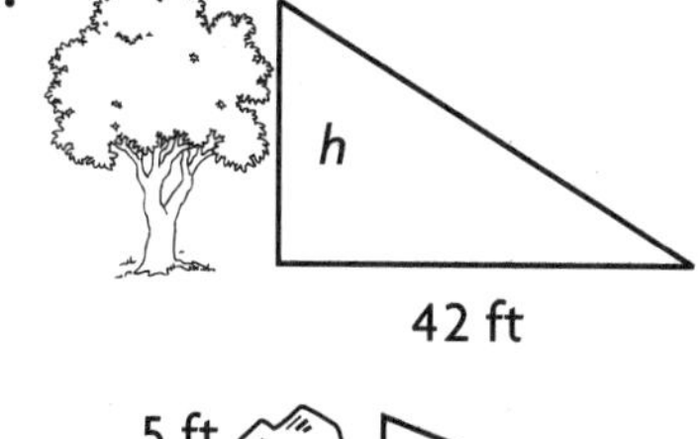

$\frac{5}{h} = \frac{7}{42}$; $h = 30$ ft

Mixed Applications

8. Tarica is trimming her parents' trees. She is 145 cm tall and can reach 40 cm above her head. She has a set of pruners that allow her to reach another 85 cm. What are the tallest branches she can trim?

   270 cm, or 2.7 m

9. A warehouse casts a shadow that is 60 ft long. At the same time, a person 5 ft tall casts a 15-ft shadow. What is the height of the warehouse?

   20 ft

Name ______________________________

Scale Drawings: Changing the Size

Vocabulary

Complete.

1. A ____scale____ is a ratio between two sets of measurements.

2. A ____scale drawing____ is a picture or diagram of a real object but is smaller or larger than the real object.

Find the ratio of length to width.

3. $l = 5$, $w = 2$ ____$\frac{5}{2}$____

4. $l = 3.4$, $w = 1.7$ ____$\frac{3.4}{1.7}$, or $\frac{2}{1}$____

Find the missing dimension.

5. scale: 1 in.:8 ft
   drawing length: 3 in.
   actual length: ____24____ ft

6. scale: 1 in.:3 ft
   drawing length: ____4____ in.
   actual length: 12 ft

7. scale: 1 cm = 15 km
   drawing length: ____9____ cm
   actual length: 135 km

8. scale: 4 cm = 1 mm
   drawing length: 1 cm
   actual length: ____0.25____ mm

Mixed Applications

9. A map shows a stone wall that is 4 in. long on the map. A building that is actually 80 ft long is drawn 2 in. long on the map. Write a proportion, and find the actual length of the wall.

   $\frac{2}{80} = \frac{4}{l}$; $l = 160$ ft

10. Darlene purchased 3 books. Each book was the same price. She paid with a $20 bill and received $6.50 in change. What was the price of each book?

    $4.50

11. A fence costs $10 per yard. A garden is 21 ft long and 18 ft wide. How much does it cost to fence in the garden?

    $260

12. A scale drawing for a barn has the scale 1 in. = 12 ft. The drawing length of the barn is 5 in. Write a proportion, and find the actual length of the barn.

    $\frac{1}{12} = \frac{5}{l}$; $l = 60$ ft

Use with text pages 368–371.

Name ____________________

Maps

Write and solve a proportion to find the actual miles. Use a map scale of 1 in. = 20 mi.

1. map distance: 4 in.

   $\frac{1}{20} = \frac{4}{n}$; 80 mi

2. map distance: 20 in.

   $\frac{1}{20} = \frac{20}{n}$; 400 mi

3. map distance: $1\frac{1}{2}$ in.

   $\frac{1}{20} = \frac{1.5}{n}$; 30 mi

4. map distance: 6 in.

   $\frac{1}{20} = \frac{6}{n}$; 120 mi

5. map distance: 18 in.

   $\frac{1}{20} = \frac{18}{n}$; 360 mi

6. map distance: $2\frac{1}{2}$ in.

   $\frac{1}{20} = \frac{2.5}{n}$; 50 mi

7. map distance: $3\frac{1}{2}$ in.

   $\frac{1}{20} = \frac{3.5}{n}$; 70 mi

8. map distance: $5\frac{1}{2}$ in.

   $\frac{1}{20} = \frac{5.5}{n}$; 110 mi

9. map distance: $12\frac{1}{2}$ in.

   $\frac{1}{20} = \frac{12.5}{n}$; 250 mi

10. map distance: 5 in.

    $\frac{1}{20} = \frac{5}{n}$; 100 mi

11. map distance: $7\frac{1}{2}$ in.

    $\frac{1}{20} = \frac{7.5}{n}$; 150 mi

12. map distance: $8\frac{1}{2}$ in.

    $\frac{1}{20} = \frac{8.5}{n}$; 170 mi

13. map distance: 25 in.

    $\frac{1}{20} = \frac{25}{n}$; 500 mi

14. map distance: $9\frac{1}{2}$ in.

    $\frac{1}{20} = \frac{9.5}{n}$; 190 mi

Mixed Applications

15. On a map of her town, Louisa notices that it is 5 in. from the school to the town hall. The scale of the map is 1 in. = $\frac{1}{2}$ mi. How far is it from school to the town hall?

    $2\frac{1}{2}$ mi

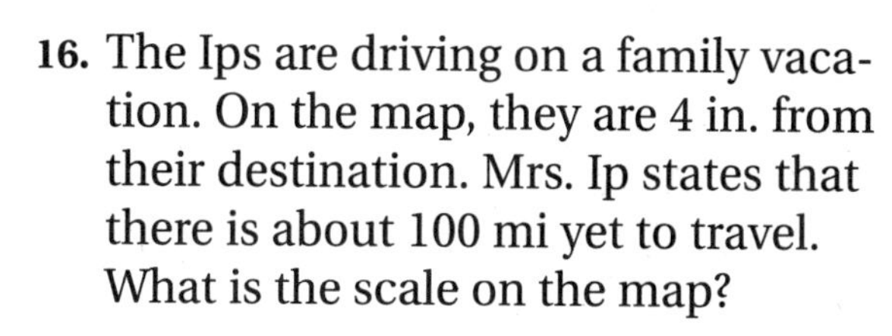

16. The Ips are driving on a family vacation. On the map, they are 4 in. from their destination. Mrs. Ip states that there is about 100 mi yet to travel. What is the scale on the map?

    1 in. = 25 mi

17. Inga makes $4.50 an hour babysitting after school. She works from 4 p.m. to 7 p.m., Tuesday through Friday. How much does she earn in a week?

    $54.00

18. A large outdoor sculpture is in the shape of a cube. The cube cost $959.85 in material to build. The artist was paid $5,000, and it cost $350 to transport it to the town square. How much did the cube project cost in all?

    $6,309.85

Name ______________________________

LESSON 20.3

Problem-Solving Strategy

Draw a Diagram

Draw a diagram to solve.

1. Bill's grandfather takes a walk every evening. He goes 3 blocks east, 4 blocks south, 3 blocks west, then 4 blocks north. Bill drew a map of the route, using the scale 1 in. = 2 blocks. Show what Bill's map looks like.

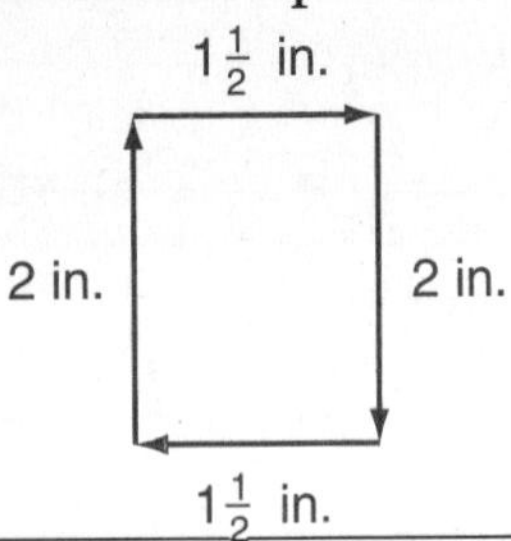

2. Marty hikes 2 mi south, 1 mi east, and 2 mi north. How far is she from her starting point? In what direction does she go to return? Draw a map with the scale 1 in. = 1 mi.

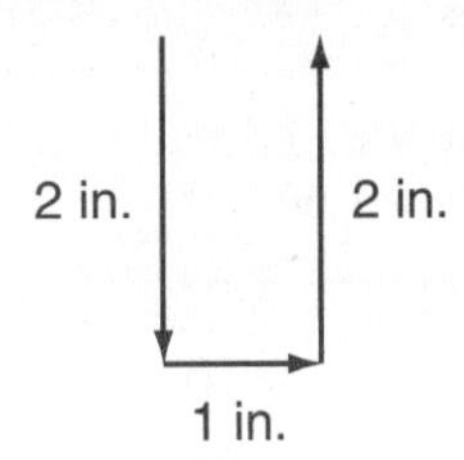

1 mi; west

Mixed Applications

Solve.

CHOOSE A STRATEGY

- Write an Equation
- Draw a Diagram
- Act It Out
- Guess and Check
- Account for All Possibilities

Choices of strategies will vary.

3. The students at Birch Middle School are going on a class trip by bus. There are 480 students in all, with 40 students on each bus. How many buses are used?

   12 buses

4. In Mrs. Hamilton's sixth-grade class, all the students are either 11 or 12 years old. There are 3 times as many 12-year-olds as 11-year-olds. There are 28 students in all. How many are 11 years old?

   7 students

5. A square playground covers 400 yd^2. What is the length of each side of the playground?

   20 yd

6. A tree casts a shadow that is 10 m long. At the same time, a nearby fence post that is 1 m high casts a 5-m shadow. What is the height of the tree?

   2 m

7. A pancake recipe calls for 1 cup pancake mix, 1 egg, 3 tbs vegetable oil, and $\frac{1}{2}$ c wheat germ. The recipe serves 4. Ty wants enough for 8. How much of each ingredient does Ty use?

   2 c mix, 2 eggs, 6 tbsp oil,
   1 c wheat germ

8. The Evanses are planning on visiting their grandparents. They want to be there by 10:30 A.M. The trip there takes 55 min. When should they leave their house?

   9:35 A.M.

Use with text pages 376–377.

Name ______________________

Golden Rectangles

Vocabulary

1. What is a Golden Ratio? a ratio equivalent to the value of about 1.6

2. What is a Golden Rectangle? a rectangle with a length-to-width ratio of about 1.6 to 1

For Exercises 3–8, tell whether the rectangle is a Golden Rectangle. Explain.

3. 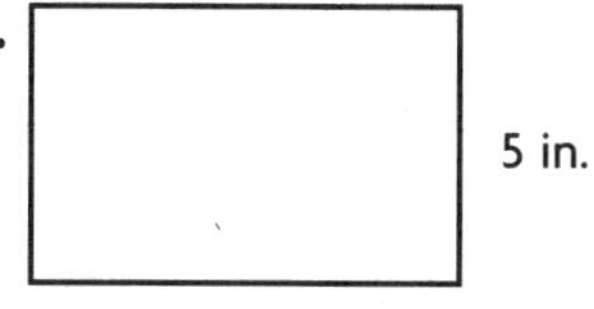

yes; $\frac{8}{5} = 1.6$

4. 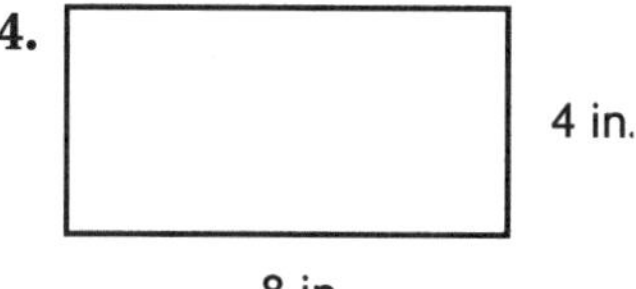

no; $\frac{8}{4} = 2.0$

5. 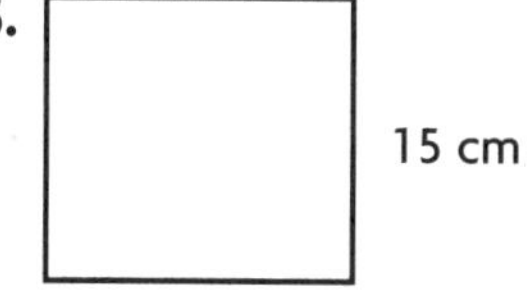

no; $\frac{17}{15} \approx 1.13$

6. 3 ft

3.5 ft

no; $\frac{3.5}{3} \approx 1.17$

7. 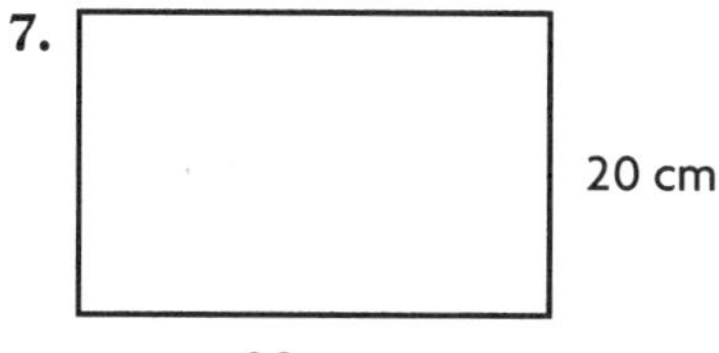

yes; $\frac{32}{20} = 1.6$

8. 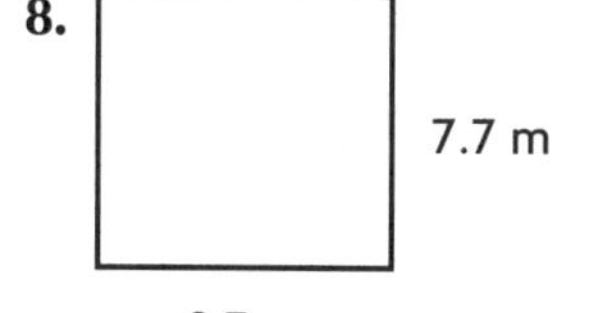

no; $\frac{8.7}{7.7} \approx 1.13$

Mixed Applications

9. The business cards for Acme Plumbing measure 8.1 cm by 5.1 cm. Is the card a Golden Rectangle? Explain.

Yes; $\frac{8.1}{5.1} \approx 1.59$, which is very close to 1.6.

10. Sean received a get-well card when he was in the hospital. The card measured 18 cm by 21 cm. Was the card a Golden Rectangle? Explain.

No; $\frac{21}{18} \approx 1.17$, which is not very close to 1.6.

11. Baseballs cost $8.98 each. Find the total cost of 8 baseballs.

$71.84

12. Dennis walks $\frac{3}{4}$ mi to school. Going home, he takes a shortcut, so it is just $\frac{5}{8}$ mi. How far does he walk in all?

$\frac{11}{8}$, or $1\frac{3}{8}$, mi

Name ______________________

LESSON 21.1

Customary Measurements

Tell what you would multiply or divide by to change the unit.

1. fluid ounces to cups
divide by 8

2. weeks to days
multiply by 7

3. inches to feet
divide by 12

4. pounds to ounces
multiply by 16

5. pints to quarts
divide by 2

6. miles to feet
multiply by 5,280

Change to the given unit.

7. 80 fl oz = **10** c
8. 18 pt = **9** qt
9. 510 ft = **170** yd
10. 720 in. = **20** yd
11. 5 months ≈ **21** weeks
12. 6 c = **48** fl oz
13. 3 gal = **12** qt
14. 6 T = **12,000** lb
15. 32 fl oz = **4** c

Use a proportion to change to the given unit.

16. 5 mi = **26,400** ft
17. 44 qt = **11** gal
18. 10 pt = **20** c
19. 157 ft = **52** yd **1** ft
20. 220 in. = **18** ft **4** in.
21. $5\frac{2}{3}$ yd = **17** ft
22. $5\frac{1}{2}$ T = **11,000** lb
23. $6\frac{1}{4}$ ft = **75** in.
24. $7\frac{1}{2}$ ft = **$2\frac{1}{2}$** yd
25. 325 ft = **108** yd **1** ft
26. $3\frac{3}{4}$ yd = **$11\frac{1}{4}$** ft
27. 15 gal = **60** qt

Mixed Applications

28. A dressmaker used 13 yd of fabric on Monday and 252 in. of fabric on Tuesday. How many yards of fabric did the dressmaker use on those two days?

20 yd

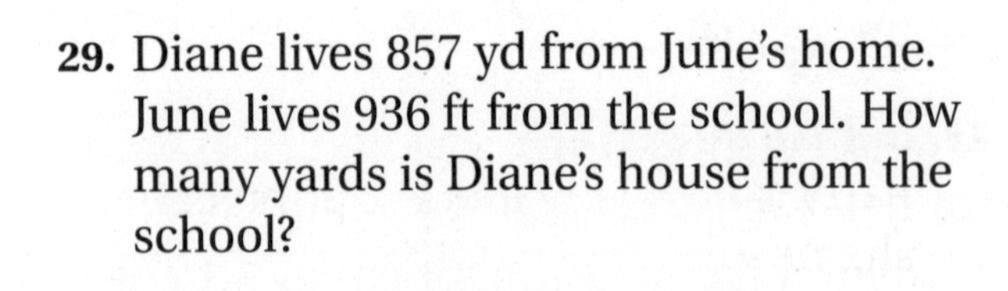

29. Diane lives 857 yd from June's home. June lives 936 ft from the school. How many yards is Diane's house from the school?

1,169 yd

30. The human body is about $\frac{3}{5}$ water by mass. What is the mass of an 85-kg adult, not including the water?

about 34 kg

31. If you roll a number cube numbered 10 to 15, what is the probability of it showing an odd number?

$\frac{1}{2}$

Use with text pages 388–389.

Name ______________________________

LESSON 21.2

Metric Measurements

Tell what you would multiply or divide by to change the unit.

1. kilograms to grams
   multiply by 1,000
2. meters to millimeters
   multiply by 1,000
3. liters to kiloliters
   divide by 1,000

Complete the pattern.

4. 1 L = 100 cL
   0.1 L = 10 cL
   0.01 L = 1 cL
5. 1,000 mg = 1 g
   100 mg = 0.1 g
   10 mg = 0.01 g
   1 mg = 0.001 g
6. 1 m = 0.001 km
   10 m = 0.01 km
   100 m = 0.1 km
   1,000 m = 1 km

Change to the given unit.

7. 40 g = 0.04 kg
8. 300 km = 300,000 m
9. 9 kL = 9,000 L
10. 6 kL = 60,000 dL
11. 300 cm = 30 dm
12. 50 dL = 500 cL

Use a proportion to change to the given unit.

13. 12 kL = 12,000 L
14. 28 g = 28,000 mg
15. 8 km = 8,000 m
16. 2.2 g = 220 cg
17. 7 dm = 0.7 m
18. 5.5 cg = 0.55 dg

Mixed Applications

19. Marcie needs 3,200 mL of punch for a party. How many liters of punch does she need?
    3.2 L
20. Teah bought 5 m of ribbon for a project. She used 36 dm of the ribbon. How much does she have left?
    14 dm
21. One year ago, a bracelet was worth $180. Its value is now 150% of that amount. How much is the bracelet worth now?
    $270
22. There are 24 students in a math class. Twelve students are wearing blue shirts and 6 students are wearing white shirts. What fraction of the students are not wearing blue or white shirts?
    $\frac{1}{4}$

Name ______________________

LESSON 21.3

Measuring One Dimension

Vocabulary

Complete.

1. The **precision** of a measurement is related to the unit of measure.

Use the ruler to measure the line segment to the given length.

2. nearest inch; nearest $\frac{1}{4}$ inch

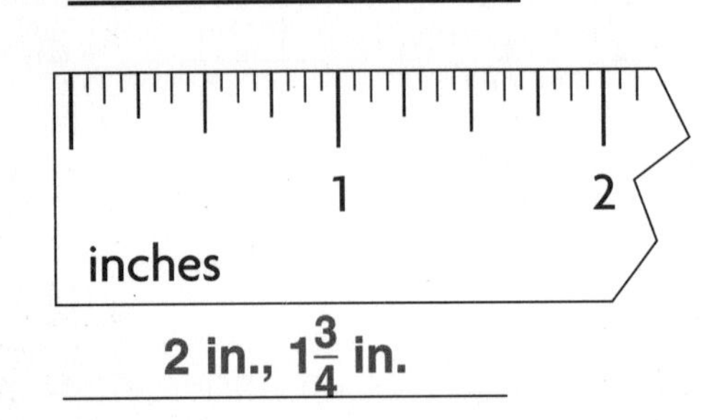

2 in., $1\frac{3}{4}$ in.

3. nearest inch; nearest $\frac{1}{2}$ inch

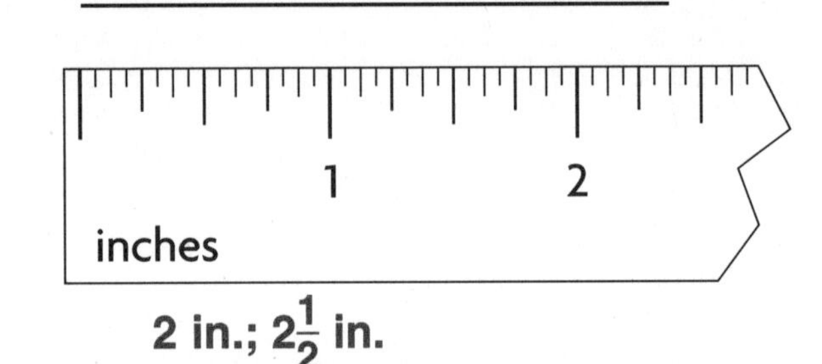

2 in.; $2\frac{1}{2}$ in.

Tell which measurement is more precise.

4. 8 cm or 82 mm

82 mm

5. 500 mm or 50 cm

500 mm

6. 9.1 cm or 9 dm

9.1 cm

7. 3 ft or 35 in.

35 in.

8. 26 in. or 2 ft

26 in.

9. 71 in. or 2 yd

71 in.

Mixed Applications

10. Joe measures the length of a couch as $6\frac{1}{2}$ ft. His sister measures it as $72\frac{3}{4}$ in. Which measurement is more precise?

$72\frac{3}{4}$ in.

11. Josh buys 2 dozen muffins. Eight are blueberry and four are pumpkin. What fraction of his purchase is either blueberry or pumpkin?

$\frac{1}{2}$

12. Kyle measures the length of a piece of wire and says it is about 4 yd. Tony measures the same wire and says it is 11 ft long. Which measurement is more precise?

11 ft

13. Thirty-two students are riding on a bus. At the first stop, 9 students get off. At the second stop, 11 students get off. What fraction of the students who got on the bus remain on the bus after the first two stops?

$\frac{3}{8}$

Use with text pages 392–393.

Name ___________________________

Networks

Vocabulary

Underline the word that best completes each sentence.

1. A network is a (graph / table) with vertices and edges.
2. You can use a network to find the (location of / distance between) two places.

Use the network to find the distance for each route. Distances are in kilometers.

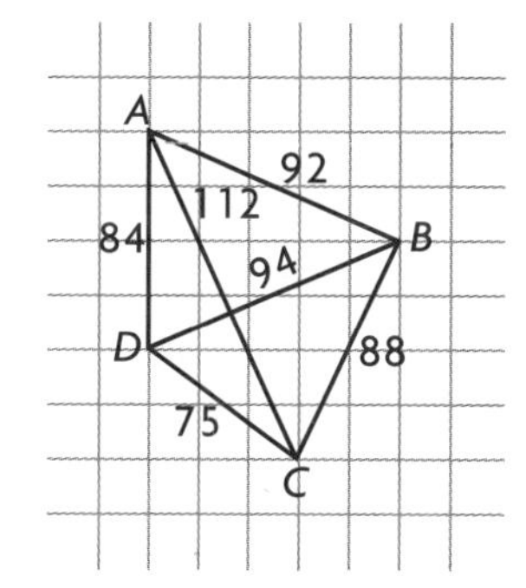

3. *ADCA*

   271 km

4. *BDAC*

   290 km

5. *ACBD*

   294 km

6. *DABC*

   264 km

Use the network to find the shorter route. Give the distance. Distances are in kilometers.

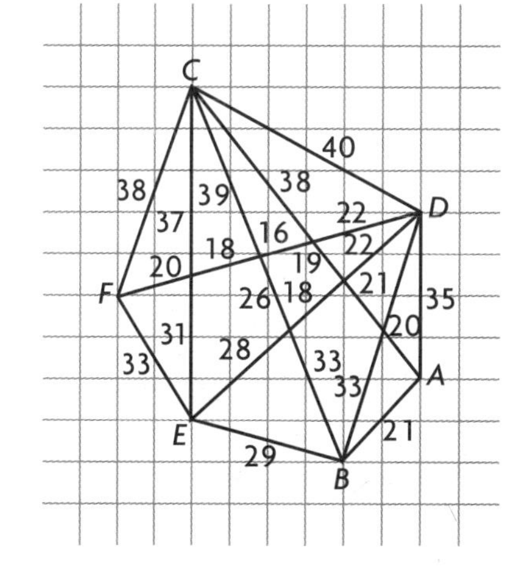

7. *DCFE* or *EFDC*

   DCFE, 111 km

8. *DBAC* or *BADC*

   BADC, 96 km

9. *ABEC* or *BACE*

   ABEC, 118 km

10. *CEDA* or *ADCE*

    ADCE, 143 km

11. *DCAB* or *CBAD*

    CBAD, 154 km

12. *ACBE* or *CABE*

    CABE, 148 km

Mixed Applications

13. What are three everyday uses of a network?

    To plan the shortest route to take when making vacation plans, preparing for a hike, or mapping out a jogging route.

14. One-third of the Ruiz family work in the family store. One-fourth of the family attend college. What fraction of the family either work in the family store or attend college?

    $\frac{7}{12}$

Name ______________________________

LESSON 21.5

Perimeter

Vocabulary

Complete.

1. The ____perimeter____ of a polygon is the distance around it.

Find the perimeter.

2.

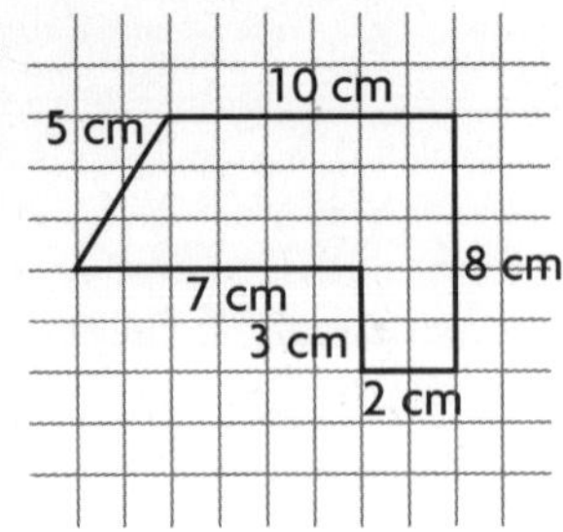

35 cm

3.

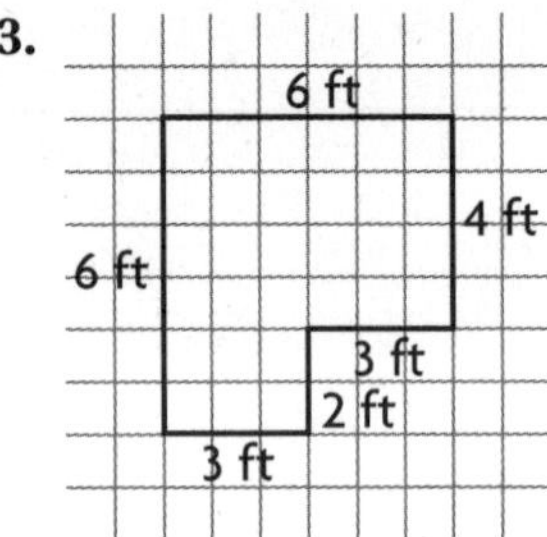

24 ft

4.

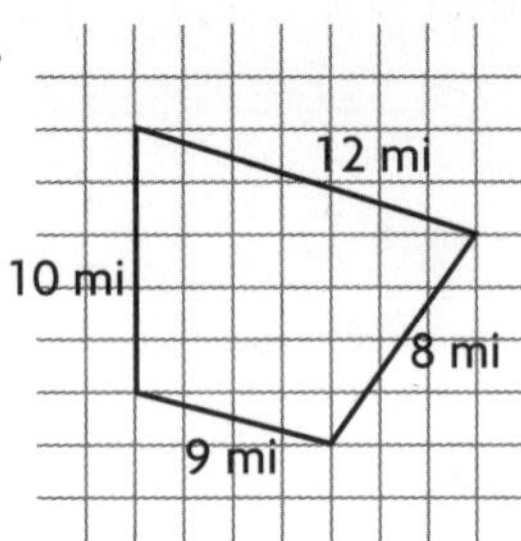

39 mi

Find the missing length.
Then find the perimeter.

5.

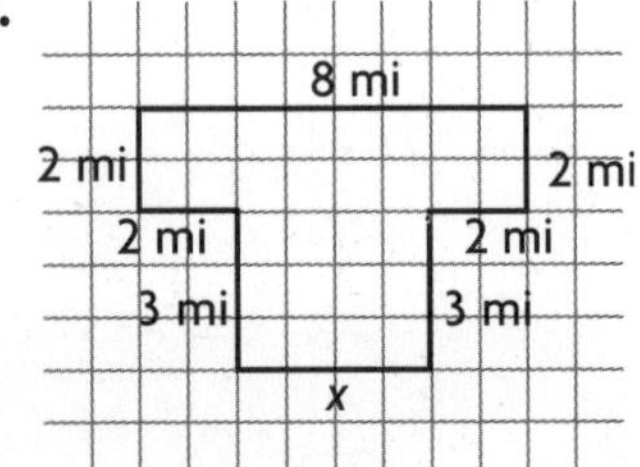

$x = 4$ mi;

26 mi

The perimeter is given.
Find the missing length.

6.

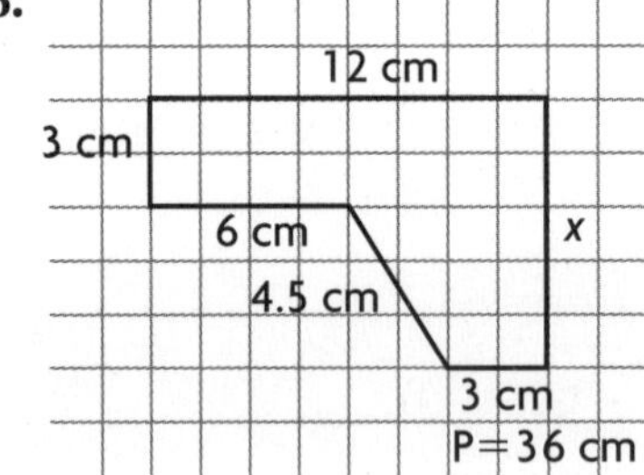

$x = 7.5$ cm

Mixed Applications

7. A rectangular flower garden is $12\frac{1}{2}$ ft long and $18\frac{2}{3}$ ft wide. What is the perimeter of the garden?

$62\frac{1}{3}$ ft

8. A comic book was worth $9.00 two years ago. Its value is now 250% of that amount. How much is the comic book worth now?

$22.50

Use with text pages 396–397.

Name ______________________

LESSON 22.1

Estimating Area

Estimate the area of the figure. Each square is 1 in.^{2} **Possible estimates are given.**

1. 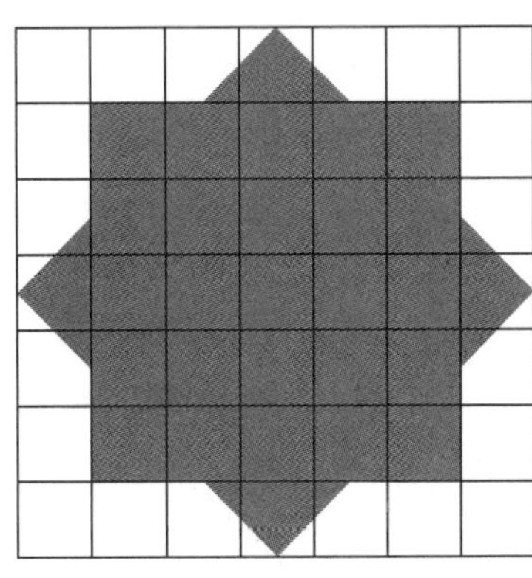

about 28 in.^{2}

2. 

about 19 in.^{2}

3. 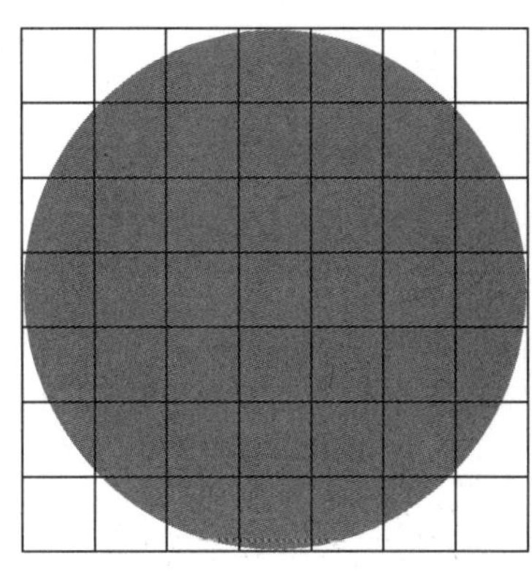

about 38 in.^{2}

4. 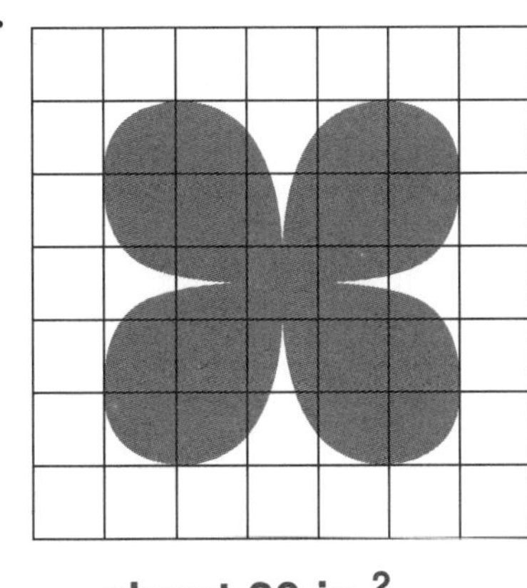

about 20 in.^{2}

5. 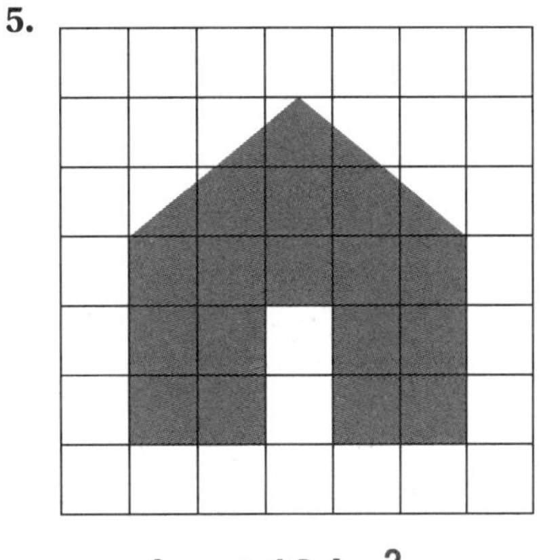

about 18 in.^{2}

6. 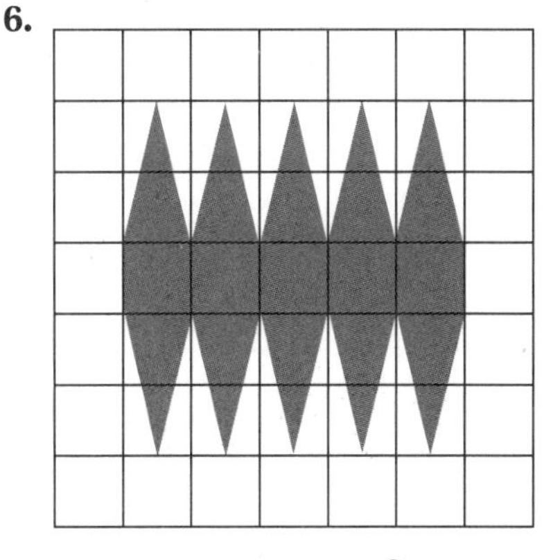

about 15 in.^{2}

Mixed Applications

7. In the diagram at the right, each square is 1 mi^2. An oil tanker ran aground and spilled oil. The shipping company needs to know the area of the oil spill to clean it up. Estimate the area.

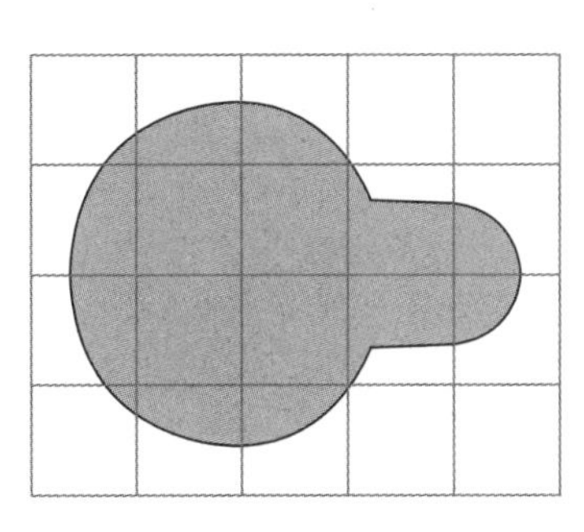

about 9 mi^2

8. Regular lawn chairs cost $34, and reclining lawn chairs cost twice as much. What will be the total cost of 2 regular chairs and 2 reclining chairs?

$204

9. Diane has 10 unpaired socks in a drawer, 5 white and 5 black. If she pulls out socks without looking, how many must she pull out to have two of the same color?

3 socks

Name ______________________________

LESSON 22.2

Problem-Solving Strategy

Use a Formula

Use a formula to solve.

1. Mr. Magruder is going to lay plastic on the cellar floor. The floor is 28 ft × 38 ft. How many square feet of plastic does he need to cover the floor?

   1,064 ft^2

2. A football stadium field is about 120 yd × 54 yd. About how many square yards of artificial turf are needed to cover the stadium field?

   about 6,400 yd^2

3. Natasha is making a bedspread. The fabric measures 3 yd × 2.5 yd. What will be the area of the bedspread? If the fabric costs $4 per square yard, how much will the bedspread cost?

   7.5 yd^2; $30

4. The Deck family is buying carpet for the living room. The living room measures 15 ft × 18 ft. How much carpet do the Decks need? If the carpet costs $4 per square foot, how much will the carpet cost?

   270 ft^2; $1,080

Mixed Applications

Solve.

CHOOSE A STRATEGY

- Work Backward • Find a Pattern • Use a Formula • Draw a Diagram • Write an Equation • Make a Chart

Choices of strategies will vary.

5. Joseph needs to be at school by 8 A.M. If the school bus takes 30 min, and Joseph needs 45 min to get ready for school, what time must Joseph get out of bed?

   6:45 A.M.

6. Nathan's front yard is 13 yd long and 7 yd wide. What is the area of his yard? If fertilizer costs $0.04 per square yard, how much will the fertilizer cost?

   91 yd^2; $3.64

7. The temperature outside rose 2°F each hour for 7 hr. Now the temperature outside is 63°F. What was the temperature 6 hours ago?

   51°F

8. Molly's garden is 18 ft long and 12 ft wide. Fencing costs $1.25 a foot. How much will it cost to fence in her garden?

   $75

9. Jolene sells cookies. In the first five months of her business, she sold 1, 2, 4, 8, and 16 dozen cookies. How many dozen cookies should Jolene expect to sell the eighth month?

   128 doz

10. Vincent travels 110 mi to visit his family. He travels 55 mi per hour. Vincent arrives at 6:30 P.M. At what time did he leave?

    4:30 P.M.

Use with text pages 406–407.

Name ____________________

LESSON 22.3

Area of Triangles and Parallelograms

Use a formula to find the area of the parallelogram.

1.

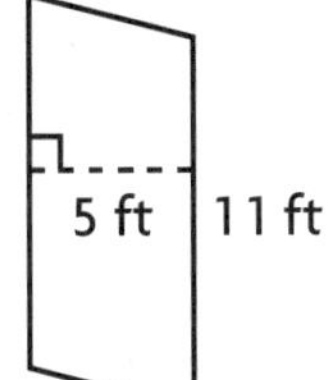

55 ft^2

2.

4.5 cm
9 cm

40.5 cm^2

3.

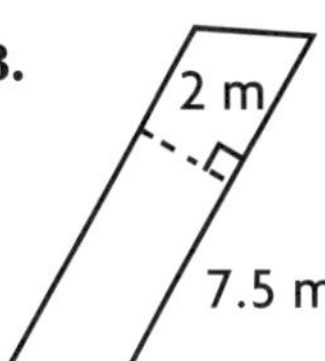

15 m^2

4. $b = 15$ in.
$h = 8$ in.

120 in.^{2}

5. $b = 4.9$ ft
$h = 3.1$ ft

15.19 ft^2

6. $b = 29\frac{3}{4}$ cm
$h = 22$ cm

$654\frac{1}{2}$ cm^2

7. $b = 3\frac{1}{2}$ m
$h = 1\frac{3}{4}$ m

$6\frac{1}{8}$ m^2

Use a formula to find the area of the triangle.

8.

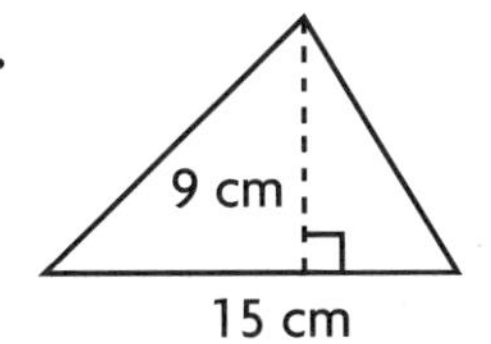

67.5 cm^2

9.

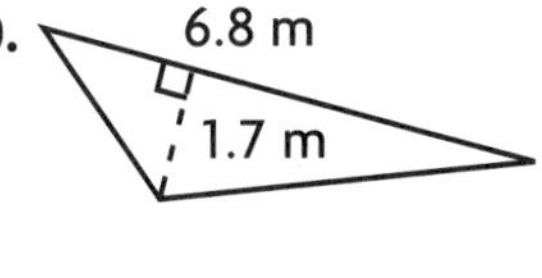

5.78 m^2

10.

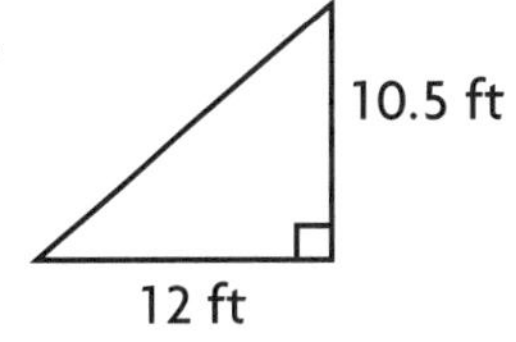

63 ft^2

11. $b = 18$ in.
$h = 8$ in.

72 in.^{2}

12. $b = 5$ ft
$h = 2\frac{1}{4}$ ft

$5\frac{5}{8}$ ft^2

13. $b = 3.2$ m
$h = 1.5$ m

2.4 m^2

14. $b = 58$ cm
$h = 29\frac{1}{8}$ cm

$844\frac{5}{8}$ cm^2

Mixed Applications

15. A pumpkin patch in the shape of a parallelogram has a base of 25 ft and a height of 16 ft. It takes 10 ft^2 to grow a large pumpkin. How many pumpkins will grow in the pumpkin patch?

40 pumpkins

16. Minerva is making a design with 6 triangles. Each triangle has a base of 6 in. and a height of 5 in. Can Minerva fit her design on an 8-in. × 11-in. sheet of paper? Explain.

no; 90 in.^{2} > 88 in.^{2}

17. How much is a 35% discount on a \$54 pair of shoes? What is the sale price of the shoes?

\$18.90; \$35.10

18. José is putting tile flooring in a 12-ft × 15-ft kitchen. How many square feet of tile does he need?

180 ft^2

Name ____________________

Changing Length and Width

Find the perimeter and area of each figure. Then double the dimensions and find the new perimeter and area.

1. 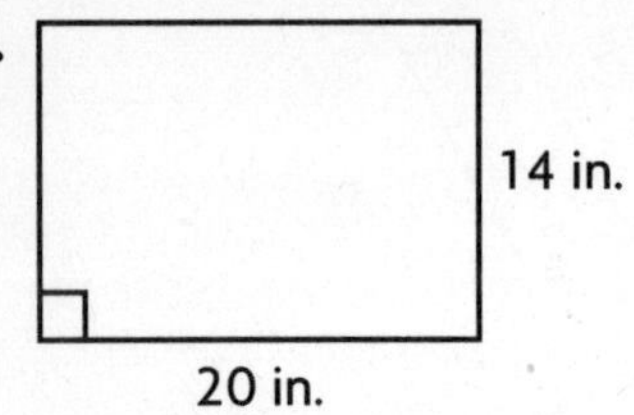

68 in., 280 in.^{2};
136 in., $1{,}120$ in.^{2}

2. 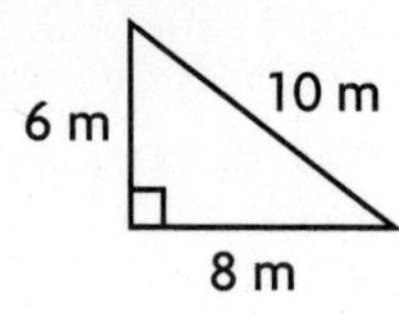

24 m, 24 m^2;
48 m, 96 m^2

3. 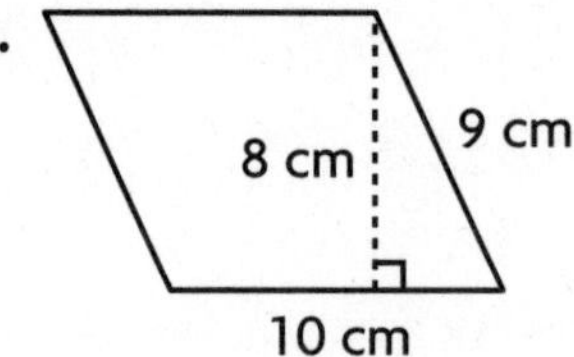

38 cm, 80 cm^2;
76 cm, 320 cm^2

Find the perimeter and area of each figure. Then halve the dimensions and find the new perimeter and area.

4. 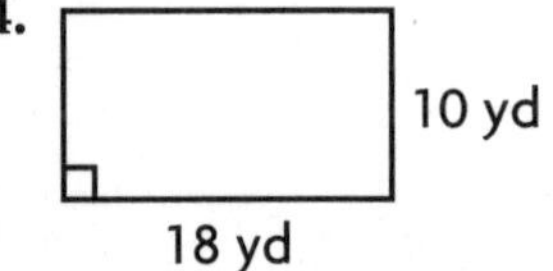

56 yd, 180 yd^2;
28 yd, 45 yd^2

5. 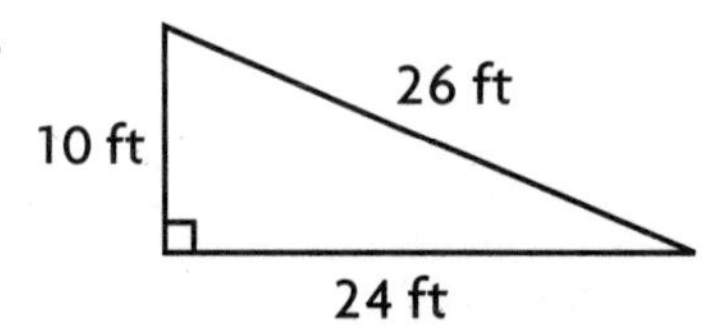

60 ft, 120 ft^2;
30 ft, 30 ft^2

6. 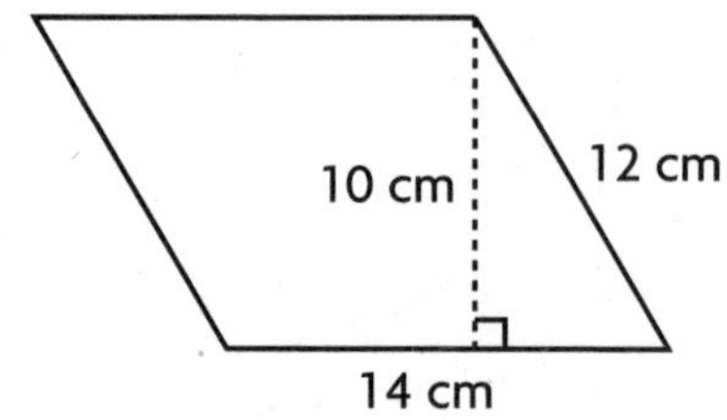

52 cm, 140 cm^2;
26 cm, 35 cm^2

Mixed Applications

7. Tyrone is putting a fence around a right-triangular garden with a base of 18 ft and a height of 24 ft. The third side is 30 ft. How much fencing does he need? What is the area of the garden?

72 ft; 216 ft^2

8. Suppose Tyrone decides to double the dimensions of the garden. How much fencing will he need? What will be the area of the garden? Compare the area of this garden to the area of the garden in Exercise 7.

144 ft; 864 ft^2; the area is 4 times as great.

9. Melissa and Amy earned \$54.40 together. Amy earned \$6.26 more than Melissa. How much did Melissa earn? How much did Amy earn?

\$24.07; \$30.33

10. A kitchen countertop is made by joining two parallelograms. Each parallelogram has a height of 24 in. The area of the countertop is 2,304 in.^{2} What is the base of each parallelogram?

48 in.

Name ______________________

Finding the Area of a Circle

Find the area of each circle. Round to the nearest whole number.
All areas are approximations.

1.
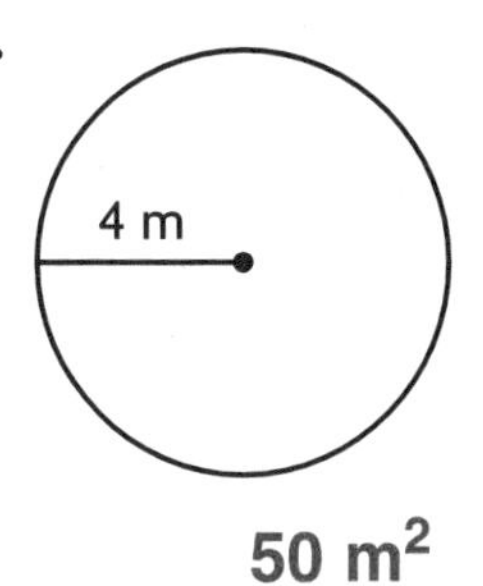

50 m^2

2.
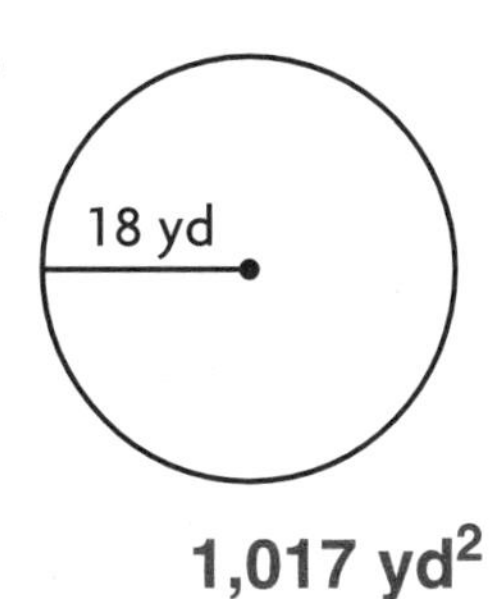

1,017 yd^2

3.
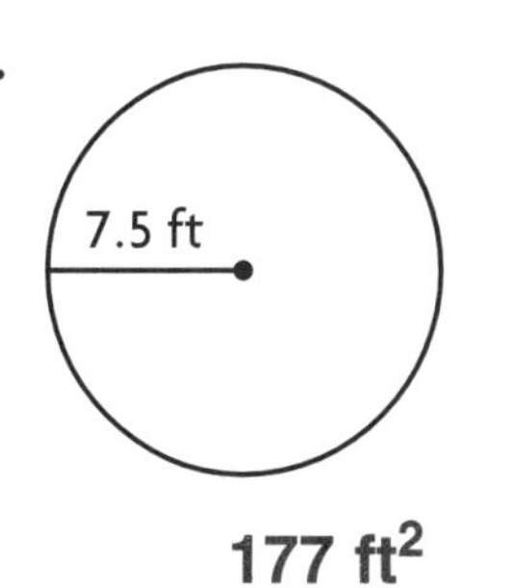

177 ft^2

Find the area of each circle. Round to the nearest tenth.

4. $r = 17$ yd 907.5 yd^2
5. $d = 38$ ft 1,133.5 ft^2
6. $r = 5.6$ m 98.5 m^2
7. $d = 10$ mm 78.5 mm^2
8. $r = 2.2$ mi 15.2 mi^2
9. $d = 54$ cm 2,289.1 cm^2
10. $r = 21$ ft 1,384.7 ft^2
11. $d = 1.8$ mi 2.5 mi^2
12. $r = 15.5$ in. 754.4 $in.^2$
13. $d = 30$ cm 706.5 cm^2
14. $r = 6.6$ yd 136.8 yd^2
15. $d = 16$ m 201.0 m^2
16. $r = 6.8$ in. 145.2 $in.^2$
17. $r = 10.2$ ft 326.7 ft^2
18. $r = 1.1$ mi 3.8 mi^2
19. $d = 9$ yd 63.6 yd^2
20. $r = 1$ mi 3.1 mi^2
21. $d = 1$ m 0.8 m^2

Mixed Applications

22. Martha grew a pumpkin with a diameter of 18 in. She wants to put it on a round tray whose area is 200 $in.^2$ Will the pumpkin fit on the tray? Explain.

No. The pumpkin needs an area of about 255 $in.^2$

23. Kenya bounces a ball. With each bounce, the ball goes half as high as on the previous bounce. On the first bounce, it bounces 96 in. How high does it bounce on the fourth bounce?

12 in.

24. A cellular phone tower receives signals from a 60-mi radius. What is the area covered by the tower's receiver?

11,304 mi^2

25. Raphael has twice as many sisters as brothers. His sister Minerva has one more brother than sister. How many girls and boys are in the family?

2 girls, 2 boys

Name ______________________________

LESSON 23.1

Estimating and Finding Volume

Vocabulary

Complete.

1. The number of cubic units needed to fill a container is called the

volume.

Find the volume of each figure.

2. 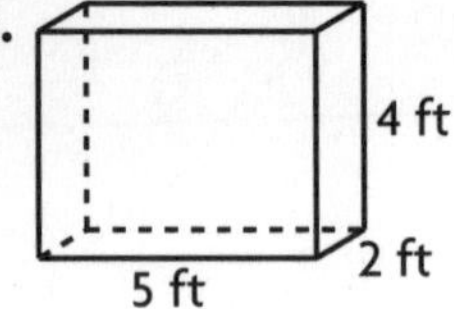

40 ft^3

3. 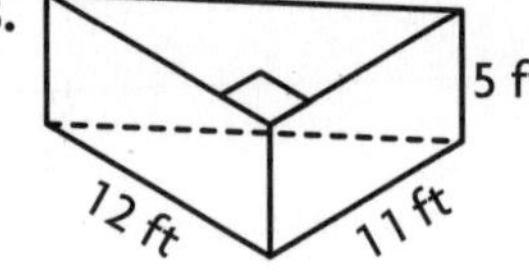

330 ft^3

4. 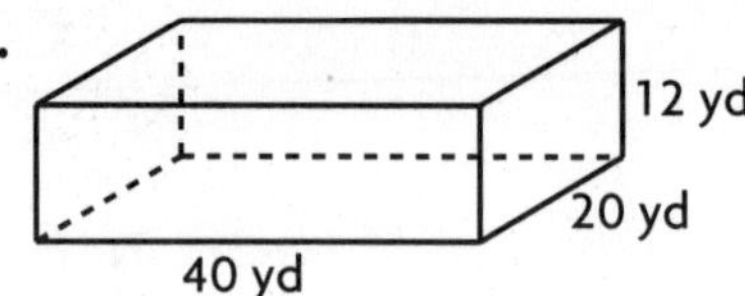

9,600 yd^3

5. 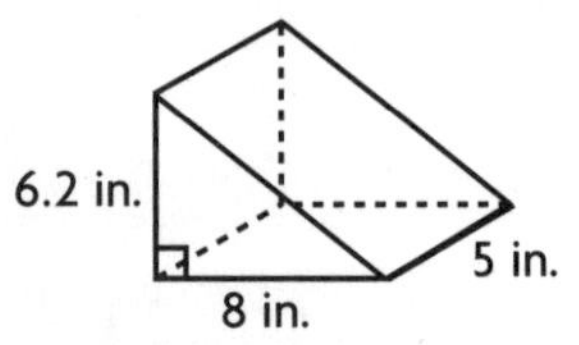

124 $in.^3$

6. 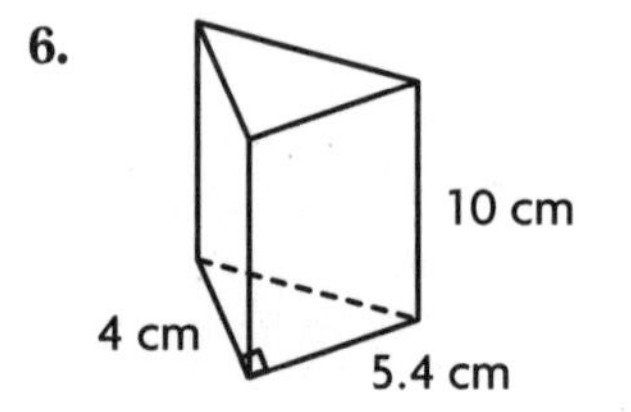

108 cm^3

7.  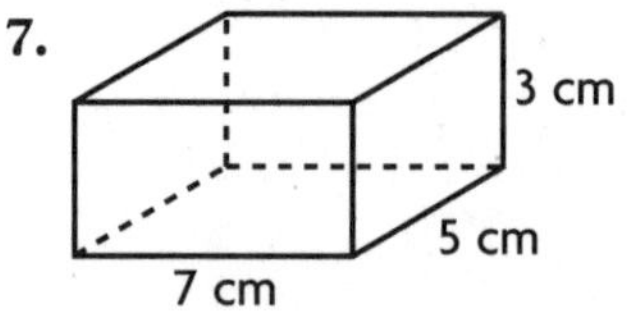

105 cm^3

Mixed Applications

8. For a party, Helen cut a block of cheese into 3 cm × 3 cm × 1 cm pieces. What is the volume of each piece?

9 cm^3

9. When Helen finished cutting, she had 36 pieces of cheese. The cheese cost $0.21 per cubic centimeter. How much did all the pieces cost?

$68.04

10. Tim wants to store his younger brother's building blocks, which are each 1 in. × 4 in. × 6 in. He has a storage box 4 in. × 12 in. × 18 in. His brother has 40 blocks. Will all the blocks fit? Explain.

No; the volume of the box is 864 $in.^3$, and the volume of 40 blocks is 960 $in.^3$

11. A cereal company is deciding on a new box design. Box A is 5 in. × 12 in. × 2 in., and Box B is 6 in. × 3 in. × 8 in. Which box will hold the most cereal? How much more cereal will it hold?

Box B; 24 $in.^3$

Use with text pages 422–425.

Name ______________________________

LESSON 23.2

Changing Length, Width, and Height

Find the volume. Then double the dimensions. Find the new volume.

1.

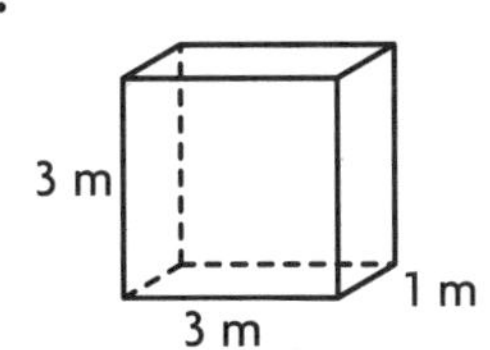

$9\ m^3$; $72\ m^3$

2.

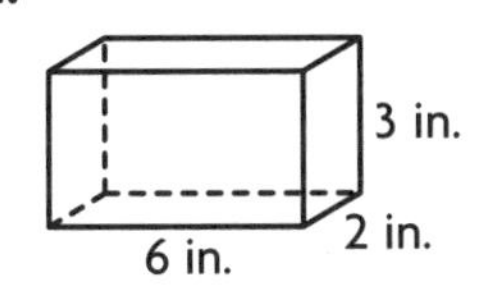

$36\ in.^3$; $288\ in.^3$

3.

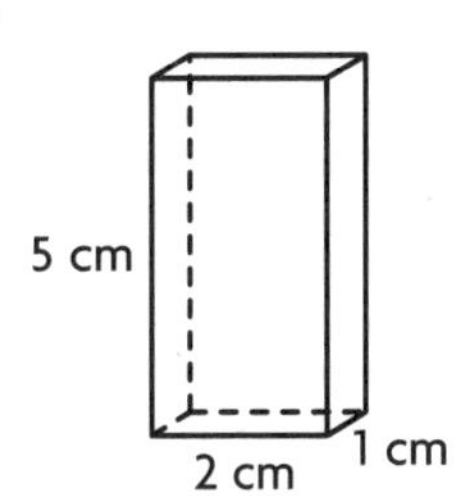

$10\ cm^3$; $80\ cm^3$

4.

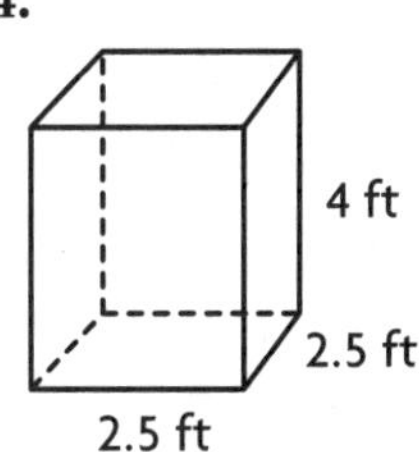

$25\ ft^3$; $200\ ft^3$

Find the volume of each prism. Then halve the dimensions and find the volume.

	Length	Width	Height	Volume	Volume (After the Dimensions Have Been Halved)
5.	5 m	4 m	2 m	$40\ m^3$	$5\ m^3$
6.	12 ft	8 ft	10 ft	$960\ ft^3$	$120\ ft^3$
7.	24 cm	3 cm	6 cm	$432\ cm^3$	$54\ cm^3$
8.	9 in.	6 in.	10 in.	$540\ in.^3$	$67.5\ in.^3$

Mixed Applications

9. Sandra has a full container of orange juice that is 6 cm × 6 cm × 12 cm. She wants to divide the juice equally between two congruent containers. The new containers are half the height, length, and width of the original container. Will each container hold half of the original volume? Explain.

No; original container: $432\ cm^3$;
new: $54\ cm^3$

10. Janet is driving to Tamars from Central City. The distance between the cities is 295 mi. She just passed a sign that said Tamars is 155 mi ahead. How far is she from Central City?

140 mi

Name ____________________

LESSON 23.3

Volume of a Cylinder

Find the volume of each figure. Round to the nearest whole number.

1. 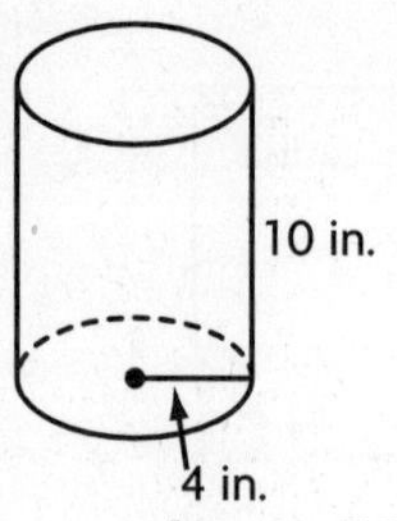

about 502 in.^{3}

2. 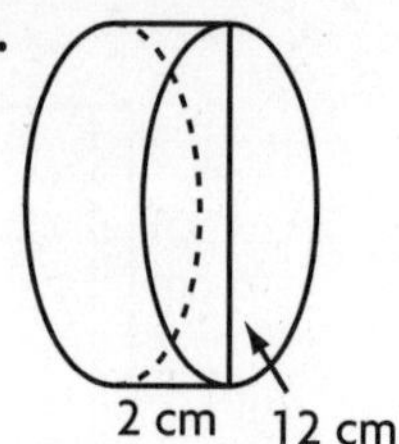

about 226 cm^3

3. 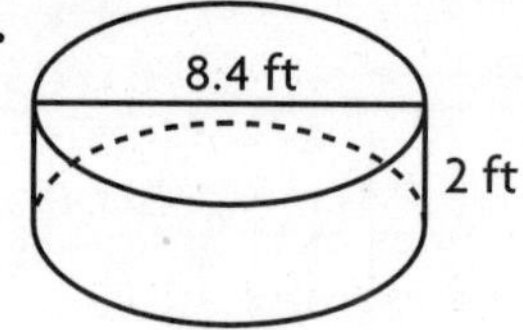

about 111 ft^3

4. 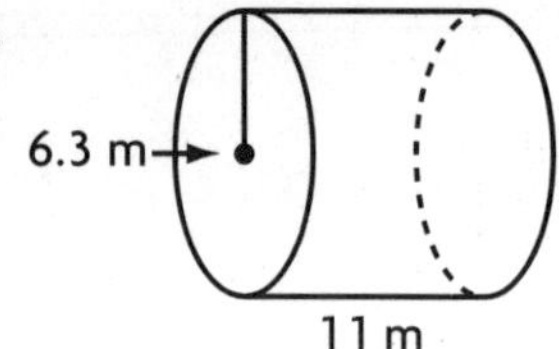

about 1,371 m^3

5. 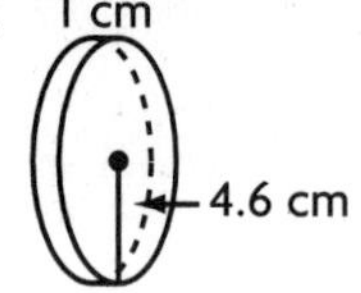

about 66 cm^3

6. 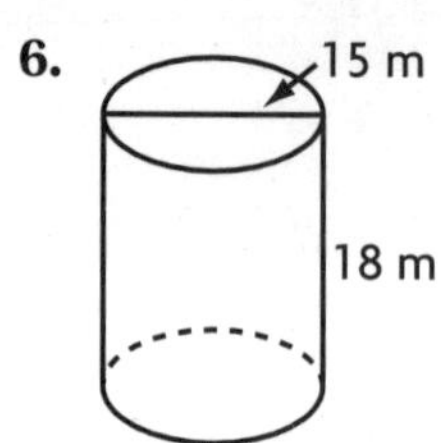

about 3,179 m^3

7. 8.5 in.
7 in.

about 1,588 in.^{3}

8. 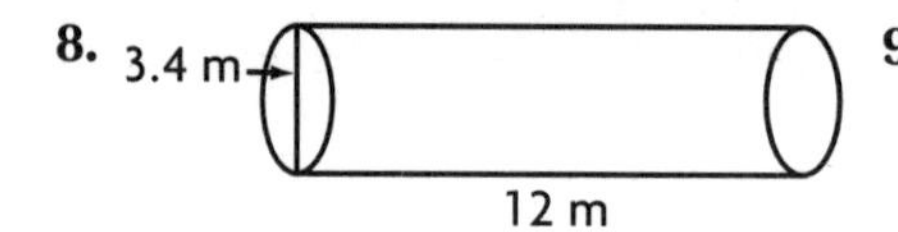

about 109 m^3

9. 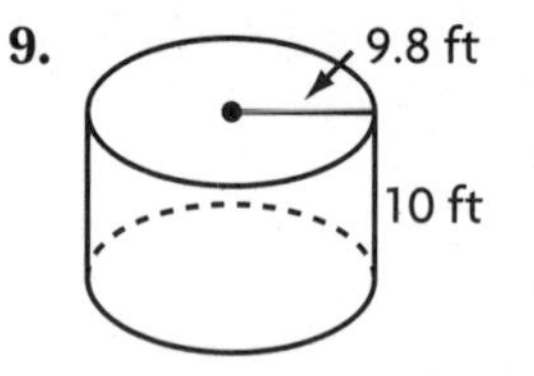

about 3,016 ft^3

Mixed Applications

10. A water storage tank is 12 meters tall and has a diameter of 20 meters. About how many cubic meters of water will the tank hold?

about 3,768 m^3

11. The tank is filled at the rate of 5 cubic meters per minute. About how many hours will it take to fill the tank?

about 13 hours

12. Jeff spent $60.00 at a bike shop. He bought a water bottle for $5.95, a lock for $23.75, and two tires that were the same price. How much did each tire cost?

$15.15

13. Danielle wants to get rid of a large juice container and pour the juice into a jar. She has 384 cubic centimeters of juice. Will a jar with a radius of 4 cm and a height of 8 cm be large enough to hold the juice?

Yes; bottle: about 402 cm^3

Use with text pages 432–433.

Name ______________________________

LESSON 23.4

Surface Area of a Rectangular Prism

Vocabulary

Complete.

1. The **surface area** is the sum of the areas of the faces of a solid figure.

Find the surface area.

2.

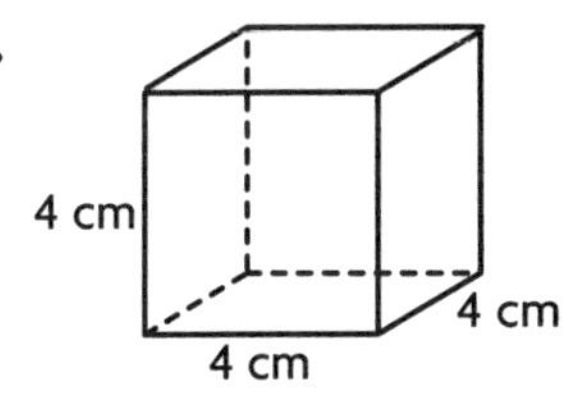

96 cm²

3.

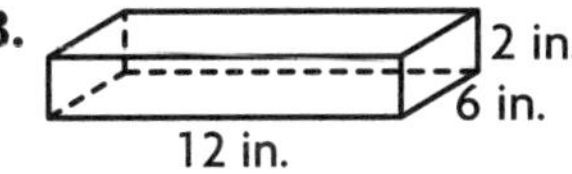

216 in.²

4.

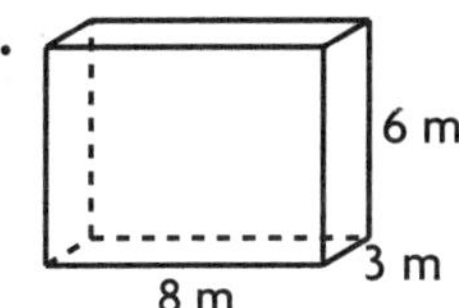

180 m²

5.

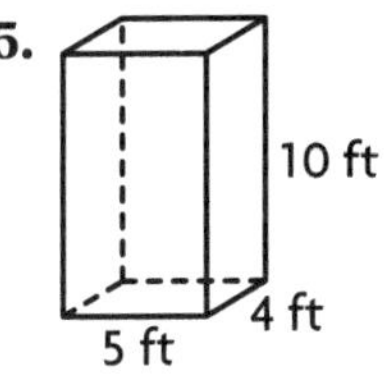

220 ft²

6.

7 m
20 m
3 m

442 m²

7.

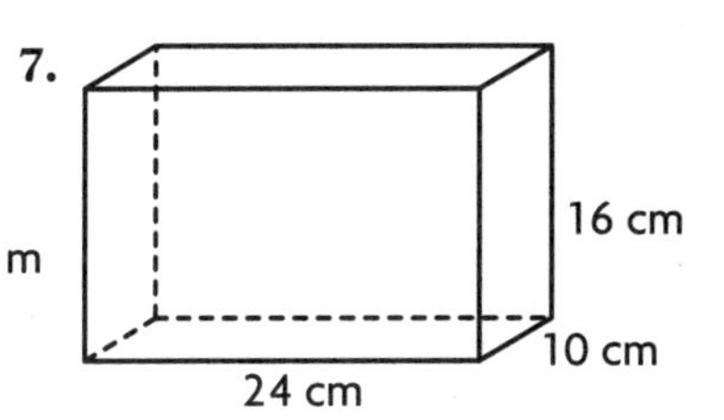

1,568 cm²

Mixed Applications

8. Jason has 2 boxes to wrap for his friend's birthday. Each box's dimensions are 20 in. × 16 in. × 12 in. How much surface area must he cover?

3,008 in.²

9. If gift wrap comes in rolls of 30 in. × 60 in., how many rolls of wrapping paper will Jason need?

2 rolls

10. What are the dimensions of the rectangular prism that is formed by this net? What is the surface area?

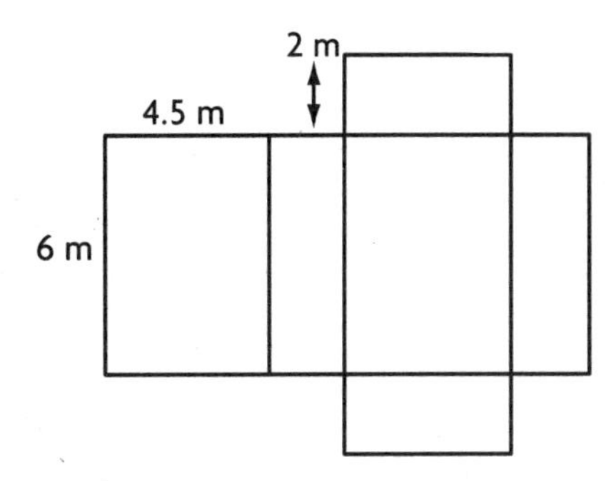

4.5 m × 6 m × 2 m; 96 m²

Name ______________________

Integers

Vocabulary

Complete.

1. **Integers** include all whole numbers, their opposites, and 0.

2. The **absolute value** of an integer is its distance from 0.

Write an integer to represent each situation. Then describe the opposite situation, and write an integer to represent it.

3. earning 7 dollars

   $^{+}7$; losing 7 dollars; $^{-}7$

4. digging a hole 2 feet deep

   $^{-}2$; making a mound 2 feet tall; $^{+}2$

5. taking 10 steps backward

   $^{-}10$; taking 10 steps forward; $^{+}10$

6. climbing up a mountain 20 feet

   $^{+}10$; going down into a cave 20 feet; $^{-}20$

Write the absolute value.

7. $|^{-}3|$ **3**
8. $|^{+}3|$ **3**
9. $|^{-}2|$ **2**
10. $|^{-}6|$ **6**
11. $|^{+}9|$ **9**
12. $|^{-}15|$ **15**
13. $|^{-}32|$ **32**
14. $|^{+}32|$ **32**
15. $|^{-}47|$ **47**
16. $|^{+}78|$ **78**
17. $|^{-}180|$ **180**
18. $|^{+}574|$ **574**

Mixed Applications

19. Del went diving with his father. They explored underwater reefs 38 feet below the surface. Write the elevation as an integer.

    $^{-}38$

20. Tina wants to wrap a present. The present is 8 in. × 6 in. × 9 in. How many square inches of wrapping paper will she need to wrap the present?

    348 sq. in.

21. The temperature outside in the morning was 15 degrees below zero. By lunchtime it was 25 degrees above zero. How many degrees did the temperature rise?

    40 degrees

22. Jill was standing outside by her favorite tree, wondering how tall it was. She noticed that the shadow of the tree was 12 feet long and her shadow was 3 feet long. Jill is 4 feet tall. How tall is the tree?

    16 feet

Use with text pages 446–447.

Name ______________________

Rational Numbers

Vocabulary

Complete.

1. Any number that can be written as the ratio $\frac{a}{b}$, where a and b are integers and $b \neq 0$, is a

   rational number.

Write each rational number in the form $\frac{a}{b}$. **Possible answers are given.**

2. $3\frac{1}{2}$ $\frac{7}{2}$
3. 0.3 $\frac{3}{10}$
4. 0.45 $\frac{45}{100}$, or $\frac{9}{20}$
5. 11.2 $\frac{112}{10}$
6. $2\frac{1}{4}$ $\frac{9}{4}$
7. 3.15 $\frac{315}{100}$
8. 15 $\frac{15}{1}$
9. 27 $\frac{27}{1}$
10. $3\frac{1}{5}$ $\frac{16}{5}$
11. 0.59 $\frac{59}{100}$
12. 370 $\frac{370}{1}$
13. $4\frac{1}{7}$ $\frac{29}{7}$

Use the Venn diagram at the right to determine in which set or sets the number belongs.

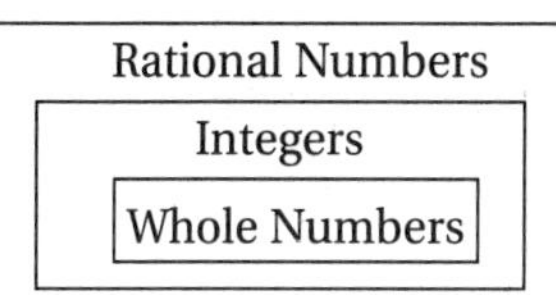

14. 1.8 R
15. $5\frac{2}{3}$ R
16. 48 all
17. $^-2$ I and R
18. 25 all
19. $^-7.1$ R
20. $\frac{1}{5}$ R
21. 0.88 R
22. 0.11 R
23. $^-1\frac{1}{8}$ R
24. $^-1$ I and R
25. 92 all

Mixed Applications

For Problems 26–27, answer *true* or *false* and give an example.

26. Every whole number is a rational number.

    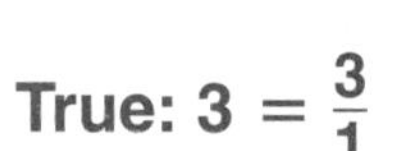
    True: $3 = \frac{3}{1}$

27. Every integer is a rational number.

    True: $\frac{-1}{2}$ is a rational number.

28. David mowed lawns for $3\frac{3}{4}$ hr. He finished at 5:00 P.M. When did he start?

    1:15 P.M.

29. Melissa scored 95 and 84 on her last two tests. What is the mean of her test scores?

    89.5

Name ____________________

LESSON 24.3

Terminating and Repeating Decimals

Vocabulary

Complete.

1. When you divide and the remainder is zero, the quotient can be written as a terminating decimal.
2. When you divide and the remainder is not zero, the quotient is a repeating decimal.

Find the terminating decimal for the fraction.

3. $\frac{3}{10}$ 0.3
4. $\frac{2}{4}$ 0.5
5. $\frac{9}{10}$ 0.9
6. $\frac{7}{25}$ 0.28
7. $\frac{11}{20}$ 0.55
8. $\frac{7}{16}$ 0.4375

Find the repeating decimal for the fraction.

9. $\frac{5}{3}$ $1.\overline{6}$
10. $\frac{4}{11}$ $0.\overline{36}$
11. $\frac{7}{30}$ $0.2\overline{3}$
12. $\frac{5}{12}$ $0.41\overline{6}$
13. $\frac{7}{15}$ $0.4\overline{6}$
14. $\frac{7}{6}$ $1.1\overline{6}$

Write the fraction as a decimal.

15. $\frac{7}{50}$ 0.14
16. $\frac{9}{8}$ 1.125
17. $\frac{13}{9}$ $1.\overline{4}$
18. $\frac{17}{20}$ 0.85
19. $\frac{11}{13}$ $3.\overline{6}$
20. $\frac{7}{4}$ 1.75

Mixed Applications

For Exercises 21–22, use the table at the right. The table shows the measurements Lisa made for a few objects and the measurements claimed by the manufacturer.

Object	Lisa	Manufacturer
Widget	$\frac{5}{8}$ in.	0.6 in.
Gizmo	$\frac{9}{16}$ in.	0.55 in.
Dial	$\frac{7}{8}$ in.	0.8 in.

21. Write as a decimal the length of the Widget Lisa measured. Is this amount different from what the manufacturer claims? If so, what is the difference?

    0.625 in.; yes; she measured 0.025 in. more.

22. Write as a decimal the length of the Gizmo Lisa measured. Is this amount different from the manufacturer's claim? If so, what is the difference?

    0.5625 in.; yes; she measured 0.0125 in. more.

23. Paula is going for a walk. She plans to walk for 3 hours. If she walks at 4 mph, how far will she walk?

    12 miles

Use with text pages 450–453.

Name ________________________________

Relationships on a Number Line

Name a rational number between the two given numbers on the number line. **Possible answers are given.**

1. 2 and $2\frac{1}{2}$ **$2\frac{1}{4}$**
2. $2\frac{1}{2}$ and 3 **$2\frac{3}{4}$**
3. 3 and $3\frac{1}{2}$ **$3\frac{1}{4}$**
4. $3\frac{1}{2}$ and 4 **$3\frac{3}{4}$**

Find a rational number between the two given numbers. **Possible answers are given.**

5. $\frac{3}{8}$ and $\frac{4}{6}$ **$\frac{11}{24}$**
6. $\frac{3}{8}$ and $\frac{2}{3}$ **$\frac{1}{2}$**
7. $1\frac{7}{8}$ and $1\frac{3}{4}$ **$1\frac{13}{16}$**
8. $^-3$ and $^-3\frac{1}{2}$ **$^-3\frac{1}{4}$**
9. 3.1 and 3.2 **3.15**
10. $^-1.7$ and $^-1.8$ **$^-1.72$**
11. $^-5.6$ and $^-5.7$ **$^-5.68$**
12. 3.04 and 3.05 **3.041**
13. $2\frac{3}{4}$ and 3 **$2\frac{7}{8}$**
14. $^-2.45$ and $^-2.46$ **$^-2.457$**
15. $\frac{3}{10}$ and 0.4 **$\frac{7}{20}$, or 0.35**
16. $\frac{2}{5}$ and 0.6 **$\frac{1}{2}$, or 0.5**
17. $2\frac{1}{4}$ and 2.5 **$2\frac{3}{8}$**
18. $^-1.09$ and $^-1.1$ **$^-1.095$**
19. 0.7 and $\frac{4}{5}$ **0.75, or $\frac{15}{20}$**
20. $^-3.5$ and $^-3.4$ **$^-3.47$**
21. 2.8 and 3 **2.88**
22. $1\frac{3}{8}$ and 1.25 **$1\frac{15}{16}$**
23. $^-5.35$ and $^-5.4$ **$^-5.37$**
24. $^-5\frac{1}{4}$ and $^-5$ **$^-5\frac{1}{8}$, or $^-5.125$**

Mixed Applications

25. Mrs. Thomas checked the gas gauge in her car and found the tank was between $\frac{1}{4}$ and $\frac{1}{2}$ full. Could the tank have been $\frac{3}{8}$ full? Explain.

    Yes; $\frac{3}{8}$ is between $\frac{1}{4}$ and $\frac{1}{2}$.

26. Fred was running a marathon. He stopped for a drink halfway between the 12.7 km mark and the 12.8 km mark. How far had he run?

    12.75 km

27. Grace cut a rectangular cake into 1-inch squares. The cake was 12 in. long by 8 in. wide. How many pieces of cake did she have?

    96 pieces

28. Jake bought baseball cards at the local swap meet. He gave 62 of them to his friends. He still had 93 left in the bag. How many baseball cards did he buy at the swap meet?

    155 cards

Name ____

LESSON 24.5

Comparing and Ordering

Use the number line to tell which number is greater.

1. $\frac{3}{4}$ or 1.75

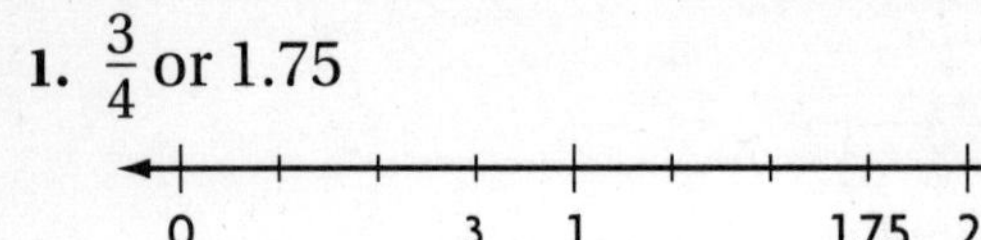

1.75

2. $-\frac{1}{4}$ or 0.5

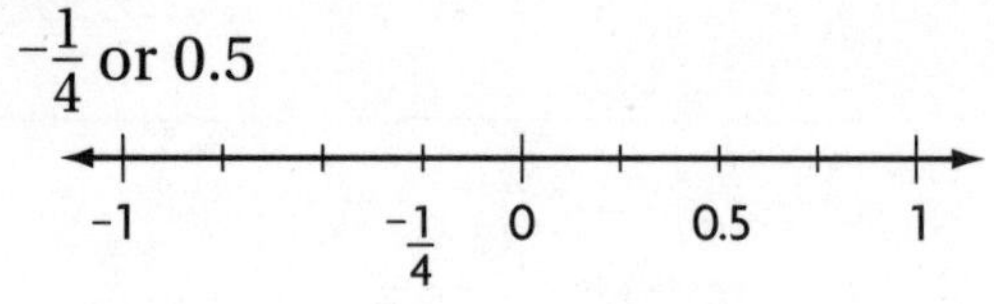

0.5

Compare. Write <, >, or = .

3. 0.4 ___<___ 0.25
4. $\frac{3}{8}$ ___>___ 0.2
5. $-2\frac{1}{5}$ ___=___ −2.2
6. $\frac{-5}{8}$ ___<___ $\frac{-3}{10}$

Compare the rational numbers and order them from least to greatest.

7. 2.9, ⁻1.7, $\frac{9}{3}$, $\frac{3}{4}$
   $^-1.7 < \frac{3}{4} < 2.9 < \frac{9}{3}$

8. $\frac{-1}{5}$, $\frac{1}{9}$, $\frac{1}{10}$, ⁻0.1
   $\frac{-1}{5} < {}^-0.1 < \frac{1}{10} < \frac{1}{9}$

9. 0, 0.8, ⁻1.4, ⁻0.6, $\frac{3}{5}$
   $^-1.4 < {}^-0.6 < 0 < \frac{3}{5} < 0.8$

10. 8.7, ⁻9.2, ⁻7.3, 6.2, $6\frac{1}{2}$, $8\frac{7}{8}$
    $^-9.2 < {}^-7.3 < 6.2 < 6\frac{1}{2} < 8.7 < 8\frac{7}{8}$

11. $4\frac{1}{4}$, $4\frac{3}{5}$, 4.9, 4.08, 0.49
    $0.49 < 4.08 < 4\frac{1}{4} < 4\frac{3}{5} < 4.9$

Compare the rational numbers and order them from greatest to least.

12. 7.3, 6, $\frac{7}{8}$, 2
    $7.3 > 6 > 2 > \frac{7}{8}$

13. 2.4, ⁻1.4, ⁻3, 4.7, 3.8
    $4.7 > 3.8 > 2.4 > {}^-1.4 > {}^-3$

14. $\frac{2}{5}$, $\frac{1}{10}$, 0.5, ⁻0.6, 0.42
    $0.5 > 0.42 > \frac{2}{5} > \frac{1}{10} > {}^-0.6$

Mixed Applications

15. Tim, Kirk, and Harry were in a race. Tim finished at 13.7 sec, Kirk at 13.65 sec, and Harry at 13.72 sec. Who took the least amount of time?

    Kirk

16. Yolanda was watching cars run around the track. The cars had practice times of $1\frac{3}{4}$ min, 1.56 min, $1\frac{4}{5}$ min, and 1.68 min. What is the longest time a car took to run a lap?

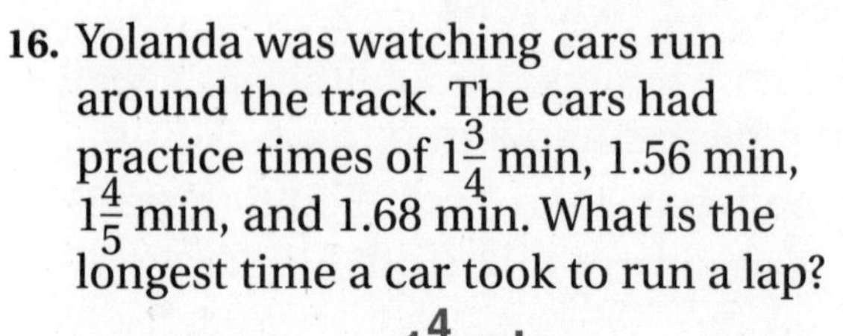

    $1\frac{4}{5}$ min

17. Jim's father built him a toy box that measures 3 ft × 4 ft × 6 ft. How much volume does Jim have to store toys?

    72 ft^3

18. Shirley went shopping with $20.00. She bought a shirt for $12.68 and a snack that cost her $4.75. How much money did she have when she got home?

    $2.57

Use with text pages 456–457.

Name ______________________

LESSON 25.1

Adding Integers

Write the addition equation modeled on the number line.

1.

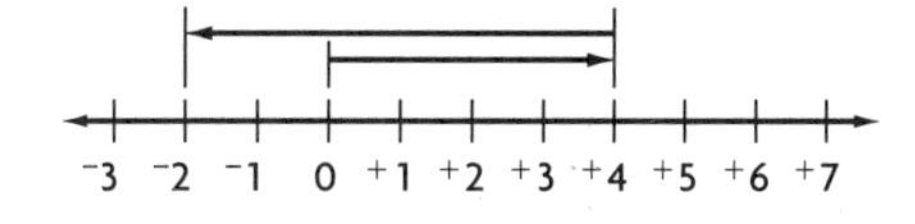

$^+4 + ^-6 = ^-2$

2.

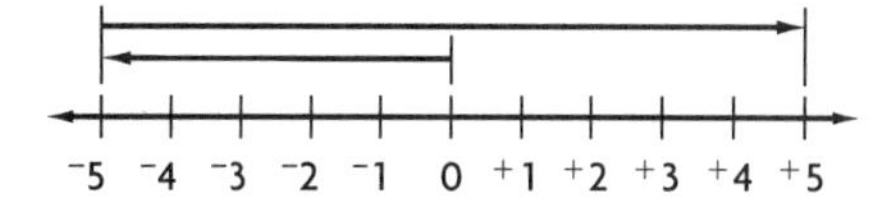

$^-5 + ^+10 = ^+5$

3.

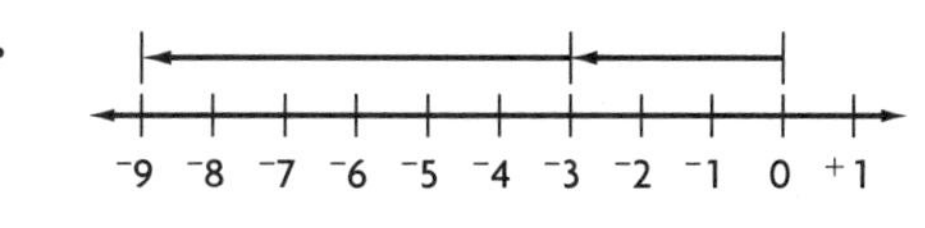

$^-3 + ^-6 = ^-9$

4.

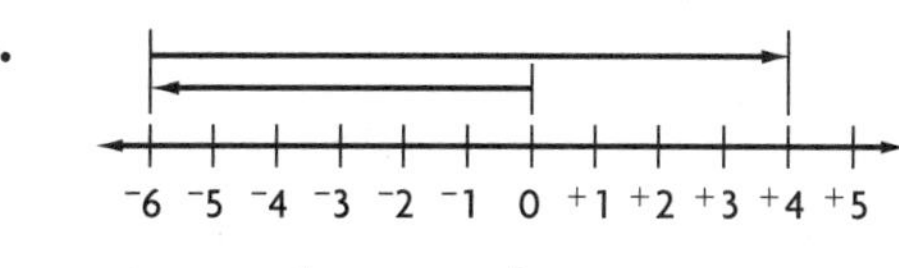

$^-6 + ^+10 = ^+4$

Find the sum.

5. $^-8 + ^-5$ **$^-13$**

6. $^+14 + ^-9$ **$^+5$**

7. $^-20 + ^-4$ **$^-24$**

8. $^+31 + ^-12$ **$^+19$**

9. $^-14 + ^-16$ **$^-30$**

10. $^+35 + ^+17$ **$^+52$**

11. $^-23 + ^-9$ **$^-32$**

12. $^+39 + ^-15$ **$^+24$**

13. $^-59 + ^-22$ **$^-81$**

14. $^+47 + ^-33$ **$^+14$**

15. $^-37 + ^-26$ **$^-63$**

16. $^+49 + ^-20$ **$^+29$**

17. $^-19 + ^-42$ **$^-61$**

18. $^+17 + ^-12$ **$^+5$**

19. $^+44 + ^-17$ **$^+27$**

20. $^-64 + ^-38$ **$^-102$**

21. $^-23 + ^+50$ **$^+27$**

22. $^-31 + ^-43$ **$^-74$**

23. $^+85 + ^-15$ **$^+70$**

24. $^-59 + ^-21$ **$^-80$**

Mixed Applications

25. Josh had an account balance of $45. He withdrew $23. Then he deposited $18. Write an addition problem to find his new balance.

$^+45 + ^-23 + ^+18 = ^+40$; $40

26. In the morning, the temperature was $^-12$°F. By noon, it had risen 7°F. What was the temperature at noon?

$^-5$°F

27. Parts for repair of a car cost $42. The labor cost for the job was $2\frac{1}{2}$ hr at $20 per hour. What was the total bill for the repairs?

$92

28. Alice charges $5 for the first hour she baby-sits and $4 for each additional hour. How much will Alice earn if she baby-sits for $3\frac{1}{2}$ hr?

$15

Name ______________________________

LESSON 25.2

Subtracting Integers

Use the number line to find the difference.

1. $^{-}6 - {}^{-}9 = {}^{-}6 + {}^{+}9 =$ **$^{+}3$**

2. $^{-}4 - {}^{+}5 = {}^{-}4 + {}^{-}5 =$ **$^{-}9$**

3. $^{-}6 - {}^{+}5 = {}^{-}6 + {}^{-}5 =$ **$^{-}11$**

4. $^{-}3 - {}^{+}7 = {}^{-}3 + {}^{-}7 =$ **$^{-}10$**

Find the difference.

5. $^{+}8 - {}^{-}9$ **$^{+}17$**
6. $^{-}14 - {}^{-}6$ **$^{-}8$**
7. $^{+}12 - {}^{-}9$ **$^{+}21$**
8. $^{+}6 - {}^{-}2$ **$^{+}8$**
9. $^{+}10 - {}^{-}3$ **$^{+}13$**
10. $^{+}11 - {}^{-}9$ **$^{+}20$**
11. $^{-}14 - {}^{-}7$ **$^{-}7$**
12. $^{-}9 - {}^{+}3$ **$^{-}12$**
13. $^{-}11 - {}^{-}9$ **$^{-}2$**
14. $^{-}9 - {}^{+}4$ **$^{-}13$**
15. $^{-}13 - {}^{+}5$ **$^{-}18$**
16. $^{-}13 - {}^{+}2$ **$^{-}15$**
17. $^{-}19 - {}^{+}7$ **$^{-}26$**
18. $^{+}16 - {}^{+}12$ **$^{+}4$**
19. $^{+}17 - {}^{-}11$ **$^{+}28$**
20. $^{-}18 - {}^{-}9$ **$^{-}9$**
21. $^{+}15 - {}^{-}14$ **$^{+}29$**
22. $^{-}19 - {}^{+}13$ **$^{-}32$**
23. $^{-}21 - {}^{+}6$ **$^{-}27$**
24. $^{-}20 - {}^{-}8$ **$^{-}12$**

Mixed Applications

25. The temperature outside is 17°, but the windchill factor is $^{-}5°$. What is the difference between the actual temperature and the windchill factor?

22°

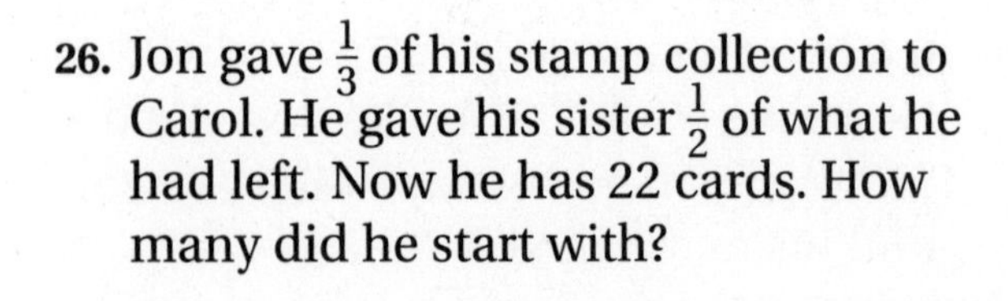

26. Jon gave $\frac{1}{3}$ of his stamp collection to Carol. He gave his sister $\frac{1}{2}$ of what he had left. Now he has 22 cards. How many did he start with?

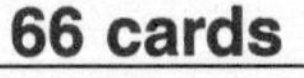

66 cards

27. The temperature at 9:00 A.M. was 21°. By noon, the temperature had decreased 5°. What is the temperature at noon?

16°

28. Rita has 9 coins. The value of the coins is $1.57. What are the coins?

6 quarters, 1 nickel, 2 pennies

Use with text pages 468–469.

Name ______________________

LESSON 25.3

Multiplying Integers

Complete the pattern.

1.	2.	3.
$^-7 \times {}^+4 = {}^-28$	$^-4 \times {}^+4 = {}^-16$	$^-9 \times {}^+4 =$ **$^-36$**
$^-7 \times {}^+3 = {}^-21$	$^-4 \times {}^+3 = {}^-12$	$^-9 \times {}^+3 = {}^-27$
$^-7 \times {}^+2 = {}^-14$	$^-4 \times {}^+2 = {}^-8$	$^-9 \times {}^+2 =$ **$^-18$**
$^-7 \times {}^+1 = {}^-7$	$^-4 \times {}^+1 = {}^-4$	$^-9 \times {}^+1 = {}^-9$
$^-7 \times 0 = 0$	$^-4 \times 0 =$ **0**	$^-9 \times 0 =$ **0**
$^-7 \times {}^-1 = {}^+7$	$^-4 \times {}^-1 = {}^+4$	$^-9 \times {}^-1 = {}^+9$
$^-7 \times {}^-2 =$ **$^+14$**	$^-4 \times {}^-2 =$ **$^+8$**	$^-9 \times {}^-2 =$ **$^+18$**
$^-7 \times {}^-3 =$ **$^+21$**	$^-4 \times {}^-3 =$ **$^+12$**	$^-9 \times {}^-3 = {}^+27$

Find the product.

4. $^-4 \times {}^+9$	5. $^+12 \times {}^-3$	6. $^-3 \times {}^-8$	7. $^+5 \times {}^-5$
$^-36$	**$^-36$**	**$^+24$**	**$^-25$**
8. $^+8 \times {}^-2$	9. $^-6 \times {}^-9$	10. $^+3 \times {}^-11$	11. $^-10 \times {}^-10$
$^-16$	**$^+54$**	**$^-33$**	**$^+100$**
12. $^-20 \times {}^-4$	13. $^+14 \times {}^-7$	14. $^-25 \times {}^+4$	15. $^+2 \times {}^-30$
$^+80$	**$^-98$**	**$^-100$**	**$^-60$**
16. $^+32 \times {}^-7$	17. $^-45 \times {}^-2$	18. $^+16 \times {}^-9$	19. $^-18 \times {}^-5$
$^-224$	**$^+90$**	**$^-144$**	**$^+90$**

Mixed Applications

20. Evaporation causes the height of water in a pool to change by $^-5$ cm each day. Write the change in the water's height over one week as a negative number.

$^-35$ cm

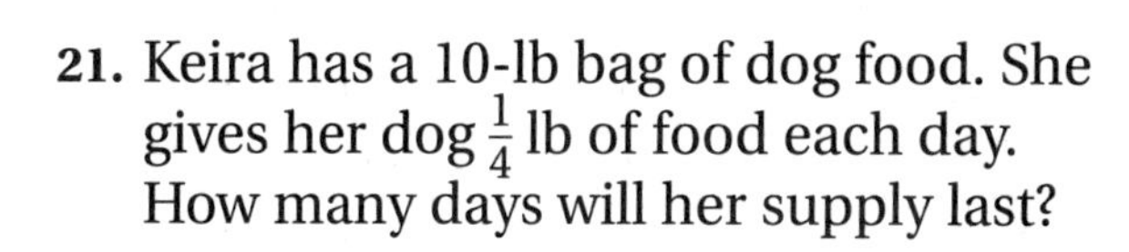

21. Keira has a 10-lb bag of dog food. She gives her dog $\frac{1}{4}$ lb of food each day. How many days will her supply last?

40 days

22. At noon the temperature was 68°. Then the temperature fell 3° every hour until midnight. Write the total change in temperature as a negative number. What was the temperature at midnight?

$^-36°$; 32°

23. There are 36 boys and girls in the parade. There are three times as many girls as boys. How many girls are in the parade?

27 girls

Name ______________________

LESSON 25.4

Dividing Integers

Write *positive* or *negative* for each quotient.

1. $^-10 \div {}^+5$ **negative**
2. $^+36 \div {}^-9$ **negative**
3. $^-44 \div {}^-11$ **positive**
4. $^+50 \div {}^-2$ **negative**

Find the quotient.

5. $^-12 \div {}^+4$ **$^-3$**
6. $^+35 \div {}^-7$ **$^-5$**
7. $^-44 \div {}^-4$ **$^+11$**
8. $^+50 \div {}^-5$ **$^-10$**
9. $^+18 \div {}^-3$ **$^-6$**
10. $-42 \div {}^-7$ **$^+6$**
11. $^+45 \div {}^-5$ **$^-9$**
12. $^+15 \div {}^+3$ **$^+5$**
13. $^-24 \div {}^-8$ **$^+3$**
14. $^+21 \div {}^-3$ **$^-7$**
15. $^-60 \div {}^-6$ **$^+10$**
16. $^-32 \div {}^+8$ **$^-4$**
17. $^+55 \div {}^-5$ **$^-11$**
18. $^-36 \div {}^-9$ **$^+4$**
19. $^+80 \div {}^-4$ **$^-20$**
20. $^+51 \div {}^-3$ **$^-17$**
21. $^-99 \div {}^-11$ **$^+9$**
22. $^+56 \div {}^+8$ **$^+7$**
23. $^-100 \div {}^+5$ **$^-20$**
24. $^-200 \div {}^-4$ **$^+50$**
25. $^-75 \div {}^+3$ **$^-25$**
26. $^+250 \div {}^-25$ **$^-10$**
27. $^-90 \div {}^-18$ **$^+5$**
28. $^-180 \div {}^+60$ **$^-3$**
29. $^-100 \div {}^-25$ **$^+4$**
30. $^-125 \div {}^+5$ **$^-25$**
31. $^+120 \div {}^-4$ **$^-30$**
32. $^-96 \div {}^+16$ **$^-6$**
33. $^+105 \div {}^-7$ **$^-15$**
34. $^-84 \div {}^+12$ **$^-7$**
35. $^+150 \div {}^-3$ **$^-50$**
36. $^-125 \div {}^+25$ **$^-5$**
37. $^-180 \div {}^-90$ **$^+2$**
38. $^+100 \div {}^-4$ **$^-25$**
39. $^-90 \div {}^-5$ **$^+18$**
40. $^-150 \div {}^+50$ **$^-3$**

Mixed Applications

41. The TipTop Company had expenses of $84,000 for one year. What was the average monthly expense?

$^-$$7,000

42. What is the perimeter of a garden that is 8 ft wide and 6 ft long?

28 ft

43. While dieting, Julia lost 8 lb in a month. How many pounds did she lose in one week?

$^-2$ lb

44. Tim wants to get a 95 average in math. He will have a total of 5 tests. How many total points will he need for all 5 tests?

475 points

Use with text pages 472–473.

Name ______________________

LESSON 26.1

Evaluating Expressions

Evaluate the numerical expression. Remember the order of operations.

1. $5 + 4 \times 6$ — **29**
2. $7 - 3 + 6$ — **10**
3. $^{-}24 \div 6 \times 3$ — **$^{-}$12**
4. $(12 - {}^{-}4) \div 4$ — **4**
5. $4^2 \times {}^{-}2$ — **$^{-}$32**
6. $^{-}2 \times (38 + {}^{-}26)$ — **$^{-}$24**
7. $^{-}5 \times 6 \div {}^{-}10$ — **3**
8. $3 \times {}^{-}5 + {}^{-}2$ — **$^{-}$17**

Evaluate the algebraic expression for the given value of the variable.

9. $x - 2$, for $x = 7$ — **5**
10. $y - 13$, for $y = 4$ — **$^{-}$9**
11. $28 + k$, for $k = 15$ — **43**
12. $7 + a$, for $a = {}^{-}16$ — **$^{-}$9**
13. $x^2 - {}^{-}2$, for $x = {}^{-}3$ — **11**
14. $y^3 - 2$, for $y = {}^{-}4$ — **$^{-}$66**
15. $3 + k - 9$, for $k = {}^{-}2$ — **$^{-}$8**
16. $^{-}4 + a^3$, for $a = 3$ — **23**
17. $5p$, for $p = {}^{-}1.7$ — **$^{-}$8.5**
18. $m \times {}^{-}12$, for $m = {}^{-}4$ — **48**
19. $j \div 8$, for $j = {}^{-}64$ — **$^{-}$8**
20. $^{-}36 \div y$, for $y = {}^{-}9$ — **4**
21. $(w - 5) \div 7$, $w = {}^{-}2$ — **$^{-}$1**
22. $a^2 \div ({}^{-}14 + 6)$, $a = {}^{-}4$ — **$^{-}$2**
23. $4 \times (p + {}^{-}6)$, for $p = {}^{-}9$ — **$^{-}$60**

Mixed Applications

24. The bottom of the well was 24 ft below the ground. The water tank stood 20 ft above the ground. Write and evaluate an expression to find the difference between the height of the water tank and the bottom of the well.

20 − $^{-}$24; 44 ft

25. Samantha has $47 in her checking account. To find the balance in her account after she writes a check, Samantha uses the expression $a-c$, where a is the amount in her checking account and c is the amount of the check. What is the balance if Samantha writes a check for $52?

$^{-}$$5

26. The Robinsons bought a circular sauna for their backyard. The diameter of the sauna is 6 ft and its height is 3 ft. What is the volume of the sauna to the nearest cubic ft?

85 ft^3

27. Judy earns $5.25 an hour at the CD store. How much does she earn in 4 weeks if she works 15 hr each week?

$315

Use with text pages 478–479.

Name ____________________

Solving Equations with Integers

Solve and check.

1. $x + 3 = 1$ — **$x = {}^-2$**
2. $8 = y + 14$ — **$y = {}^-6$**
3. $k + 8 = {}^-3$ — **$k = {}^-11$**
4. ${}^-5 = n + 7$ — **$n = {}^-12$**
5. $c - 6 = {}^-3$ — **$c = 3$**
6. ${}^-9 = w - 13$ — **$w = 4$**
7. $r - 5 = {}^-6$ — **$r = {}^-1$**
8. ${}^-18 = u - 11$ — **$u = {}^-7$**
9. $x + 120 = 40$ — **$x = {}^-80$**
10. ${}^-15 = y + 82$ — **$y = {}^-97$**
11. $k - 245 = {}^-184$ — **$k = 61$**
12. ${}^-456 = n - 385$ — **$n = {}^-71$**
13. ${}^-2y = 26$ — **$y = {}^-13$**
14. ${}^-6t = 42$ — **$t = {}^-7$**
15. ${}^-7p = {}^-21$ — **$p = 3$**
16. ${}^-8m = {}^-40$ — **$m = 5$**
17. $4y = {}^-24$ — **$y = {}^-6$**
18. $9d = {}^-72$ — **$d = {}^-8$**
19. $\frac{y}{2} = {}^-7$ — **$y = {}^-14$**
20. $\frac{x}{4} = {}^-19$ — **$x = {}^-76$**
21. $\frac{y}{{}^-3} = 12$ — **$y = {}^-36$**
22. $\frac{a}{{}^-5} = 14$ — **$a = {}^-70$**
23. $\frac{m}{{}^-9} = {}^-10$ — **$m = 90$**
24. $\frac{d}{{}^-11} = {}^-15$ — **$d = 165$**
25. ${}^-260 = 5p$ — **$p = {}^-52$**
26. ${}^-400 = {}^-20r$ — **$r = 20$**
27. ${}^-150 = \frac{x}{{}^-24}$ — **$x = 3{,}600$**
28. ${}^-541 = \frac{a}{12}$ — **$a = {}^-6{,}492$**
29. $\frac{m}{{}^-6} = 34$ — **$m = {}^-204$**
30. ${}^-8t = 184$ — **$t = {}^-23$**
31. $\frac{n}{{}^-15} = {}^-62$ — **$n = 930$**
32. ${}^-320 = \frac{c}{5}$ — **$c = {}^-1{,}600$**

Mixed Applications

For Problems 33–34, write an equation for the word sentence.

33. Negative four multiplied by a number, *x*, is forty-four.

    ${}^-4x = 44$

34. Eight more than a number, *x*, is negative twelve.

    $x + 8 = {}^-12$

35. Dolores went to the store with \$10.00. She bought a soda for \$0.85, a magazine for \$3.95, and a bag of chips for \$0.95. Does she have enough to buy a book for \$4.50? Explain.

    No; \$10 − \$5.75 = \$4.25, \$4.25 < \$4.50

36. The temperature at 7:00 was 12°F. Three hours later it had dropped 20°F. What is the temperature now?

    ${}^-8$°F

Use with text pages 480–481.

Name ____________________

LESSON 26.3

Inequalities

Vocabulary

Complete.

1. An ___inequality___ uses <, >, ≤, or ≥.

Find all whole-number solutions of the inequality.

2. $x < 1$ ___0___

3. $x < 6$ ___0, 1, 2, 3, 4, 5___

4. $a \leq 7$ ___0, 1, 2, 3, 4, 5, 6, 7___

Sketch a graph on a number line to show all of the solutions to the inequality.

5. $x < 9$

6. $x \geq 6$

7. $x > 14$

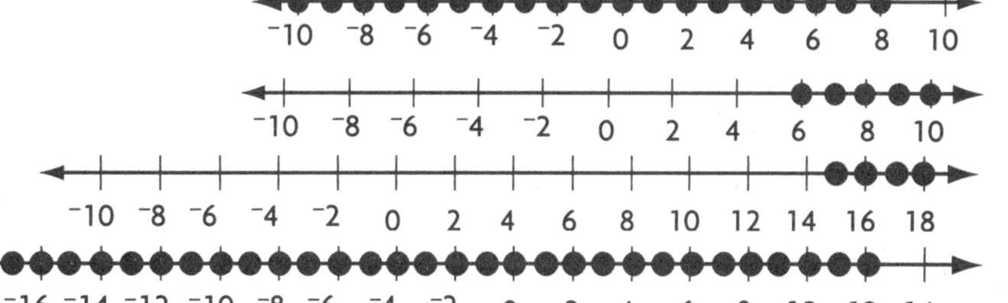

8. $x \leq 12$

Write the algebraic inequality represented by the number line.

9. 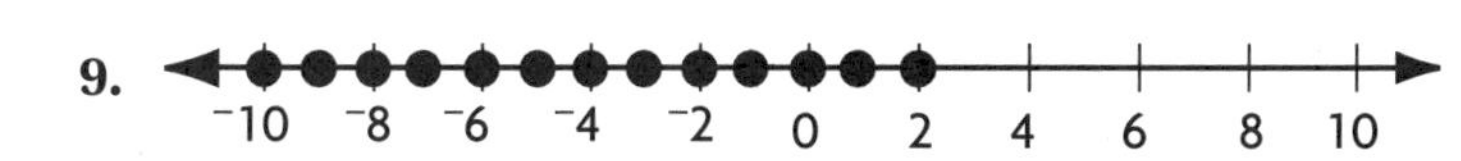 ___$x < 3$___

10. 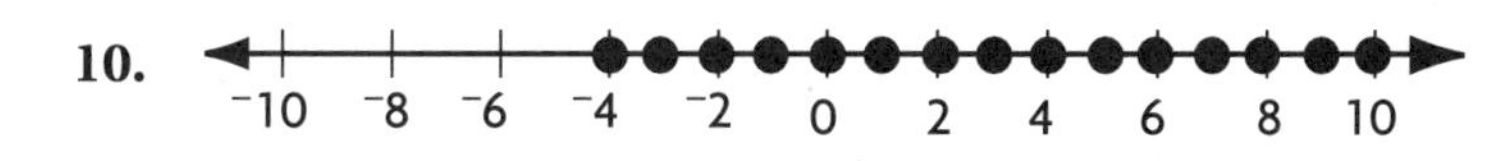 ___$x \geq {}^{-}4$___

Mixed Applications

11. Why is 3 not included as a solution in Exercise 9?

___because 3 is not part of the solution___

12. Why is ⁻4 included as a solution in Exercise 10?

___because ⁻4 is part of the solution___

13. A farmer must make certain that his zucchini are less than 6 in. long before he can ship them to market. Write an algebraic inequality representing the lengths of the zucchini he can ship.

___$x < 6$ in.___

14. Tom is older than his sister Shelly. If t represents Tom's age and s represents Shelly's age, write an algebraic inequality that shows the relationship between their ages.

___$t > s$ or $s < t$___

Name ______________________________

LESSON 26.4

Graphing on the Coordinate Plane

Vocabulary

Write the correct letter from Column 2.

Column 1

__c__ 1. divide the coordinate plane into 4 quadrants

__e__ 2. the place where the x-axis and y-axis intersect

__b__ 3. what two perpendicular lines that intersect form

__a__ 4. the numbers that go with a point on the graph

__d__ 5. a section of the coordinate plane

Column 2

a. ordered pair

b. coordinate plane

c. axes

d. quadrant

e. origin

Write the ordered pair for the point on the coordinate plane.

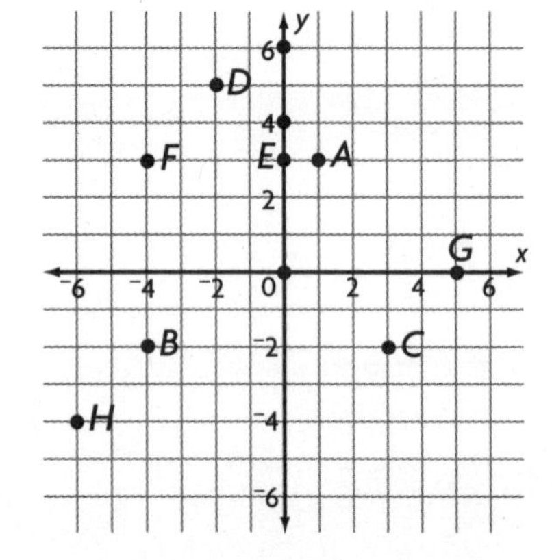

6. point A __(1,3)__
7. point B __(⁻4,⁻2)__
8. point C __(3,⁻2)__
9. point D __(⁻2,5)__
10. point E __(0,3)__
11. point F __(⁻4,3)__

Locate the point for each ordered pair on the graph. **Check students' graphs.**

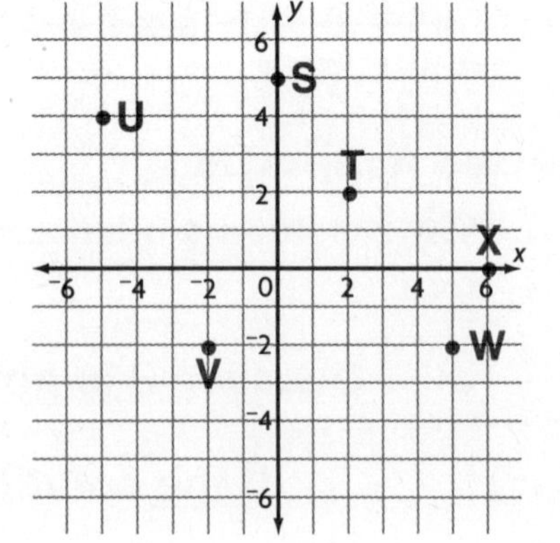

12. S (0,5)
13. T (2,2)
14. U (⁻5,4)
15. V (⁻2,⁻2)
16. W (5,⁻2)
17. X (6,0)

Mixed Applications

18. Locate the points (⁻1,4), (⁻1,⁻2), and (1,4) on the same coordinate plane. Connect the points in the listed order. What kind of geometric shape do you have?

__right triangle__

19. Dan has seven coins. The coins equal $1.15. What are the coins?

__4 quarters and 3 nickels__

Use with text pages 486–489.

Name ______________________________

Graphing Relations

Use the first three values of *x* and *y* to complete the table.

1.

INPUT (x)	1	2	3	4	5
OUTPUT (y)	3	4	5	6	7

2.

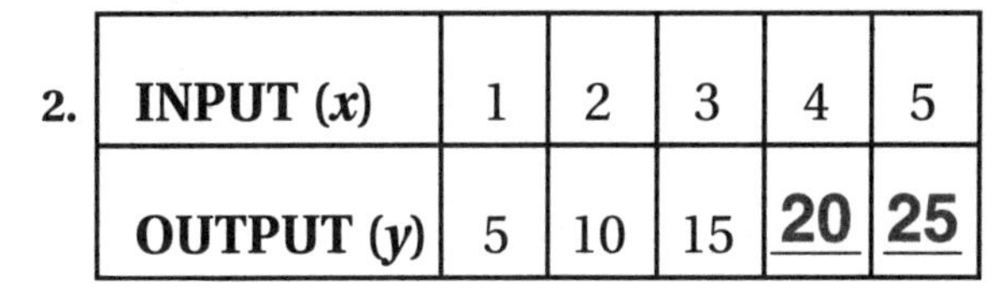

INPUT (x)	1	2	3	4	5
OUTPUT (y)	5	10	15	20	25

3. Locate the points for the ordered pairs of the relation from Exercise 1.

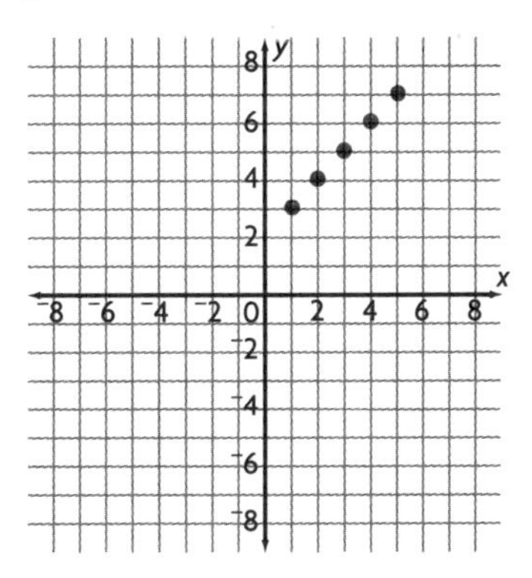

4. Locate the points for the ordered pairs of the relation from Exercise 2.

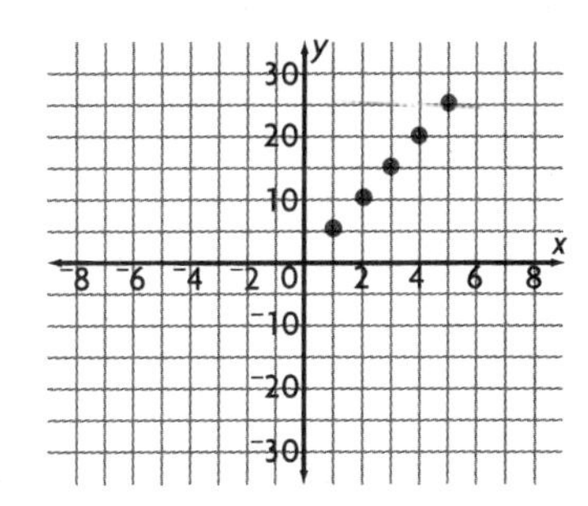

5. What expression can you write, using x, to get the value of y in Exercise 1?

$y = x + 2$

6. What expression can you write, using x, to get the value of y in Exercise 2?

$y = 5x$

Mixed Applications

7. Use the expression $x + 6$ to make an input-output table. Use the whole numbers from 0 to 4 as the input, x. Find the output, y. Locate the points for the ordered pairs on a coordinate plane.

INPUT (x)	0	1	2	3	4
OUTPUT (y)	6	7	8	9	10

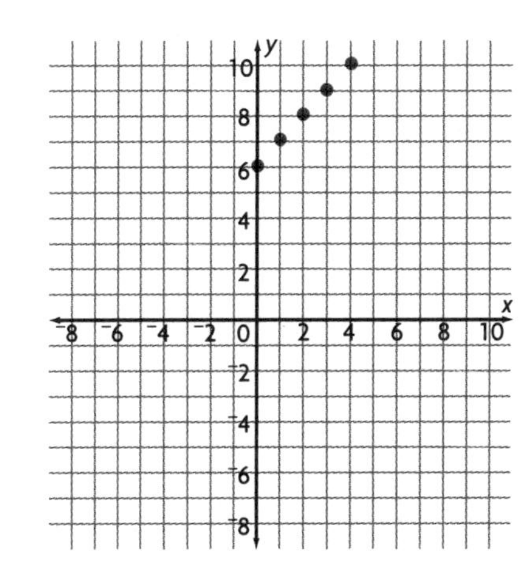

8. Write an algebraic inequality for the word sentence: All numbers x are greater than or equal to negative four.

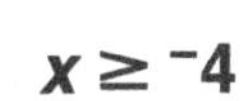
$x \geq {}^{-}4$

Use with text pages 490–491.

Name ____________________

Transformations on a Coordinate Plane

In Exercise 1, translate the figure 4 units to the right and 3 units up. In Exercises 2–3, reflect the figure across the *x*-axis.

1.

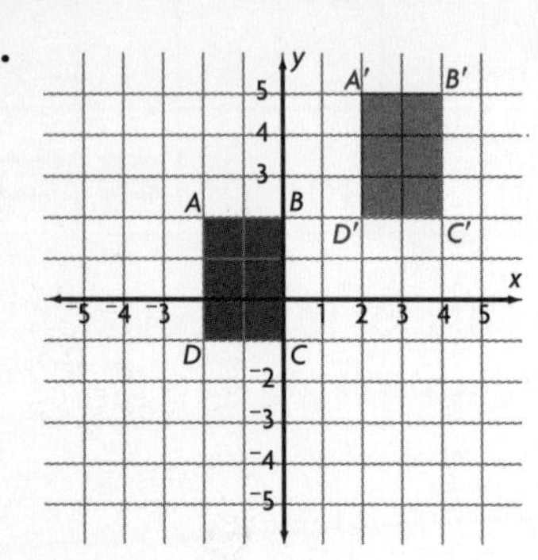

2.

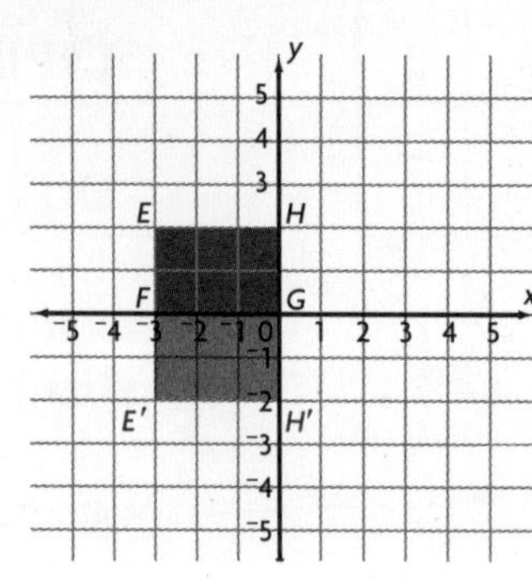

3.

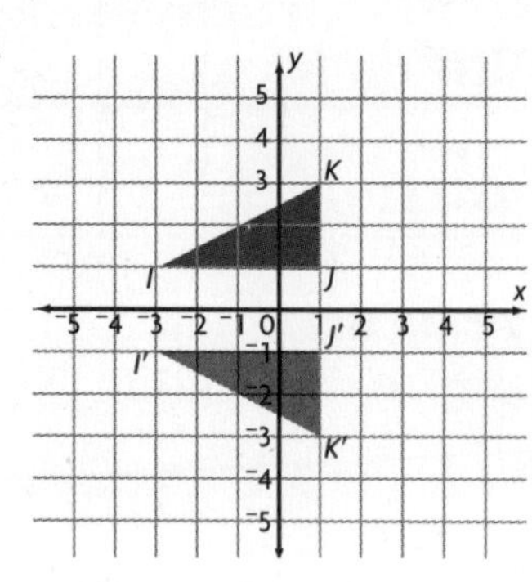

Rotate the figure around the origin according to the directions given. Give the new coordinates.

4. 90° clockwise

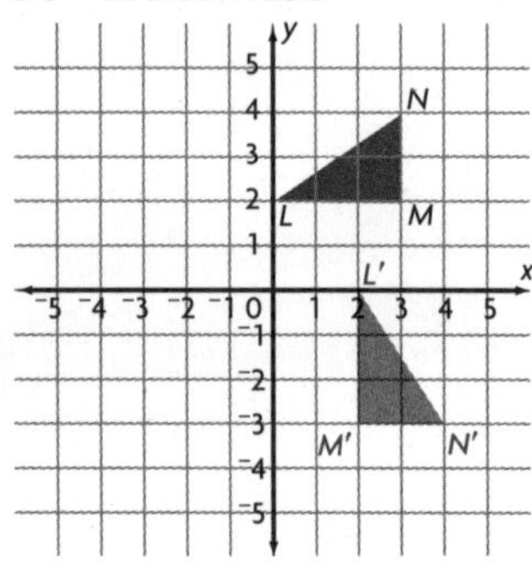

$L'(2,0)$, $M'(2,{}^-3)$, $N'(4,{}^-3)$

5. 180° counterclockwise

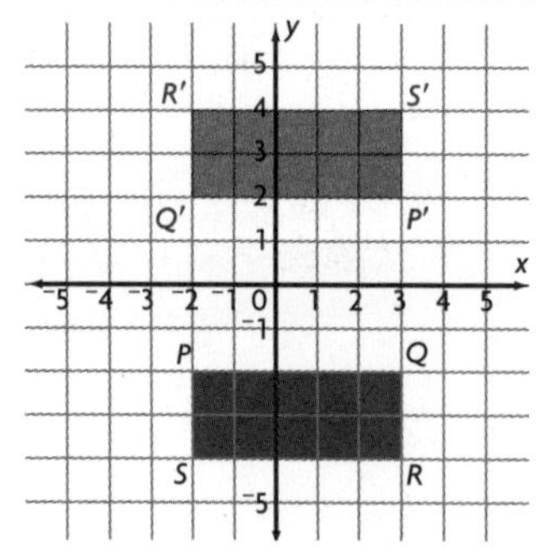

$P'(3,2)$, $Q'({}^-2,2)$, $R'({}^-2,4)$, $S'(3,4)$

6. 90° counterclockwise

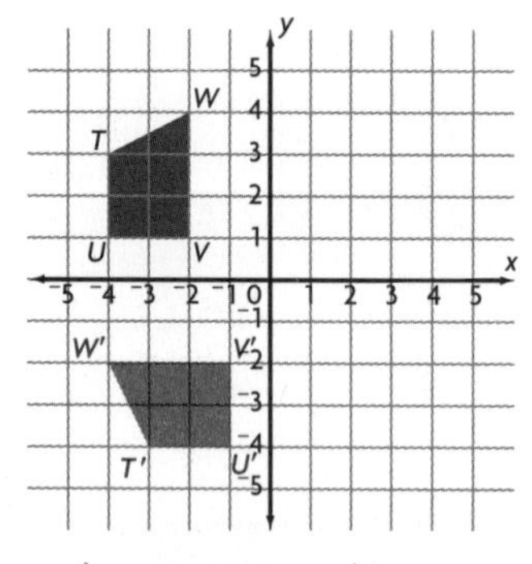

$U'({}^-1,{}^-4)$, $V'({}^-1,{}^-2)$, $W'({}^-4,{}^-2)$, $T'({}^-3,{}^-4)$

Mixed Applications

7. The coordinates of rectangle *PQRS* are $({}^-3,1)$, $({}^-3,{}^-3)$, $(4,1)$, and $(4,{}^-3)$. If you translate rectangle *PQRS* 2 units to the right, what are the new coordinates?

$({}^-1,1)$, $({}^-1,{}^-3)$, $(6,1)$, and $(6,{}^-3)$

8. One night, the temperature fell 2°F per hour between midnight and 6 A.M. The temperature was 27°F at midnight. Find the temperature at 6 A.M.

15°F

9. Mrs. Ruiz needs enough coffee for 12 people to have 2 cups each. How many quarts of coffee should she make?

6 qt

10. Ron has a total of 45 ducks and pigs. The animals have a total of 144 feet. How many ducks does Ron have?

18 ducks

Use with text pages 500–503.

Name ______________________________

LESSON 27.2

Problem-Solving Strategy

Finding Patterns on the Coordinate Plane

Find a pattern and solve.

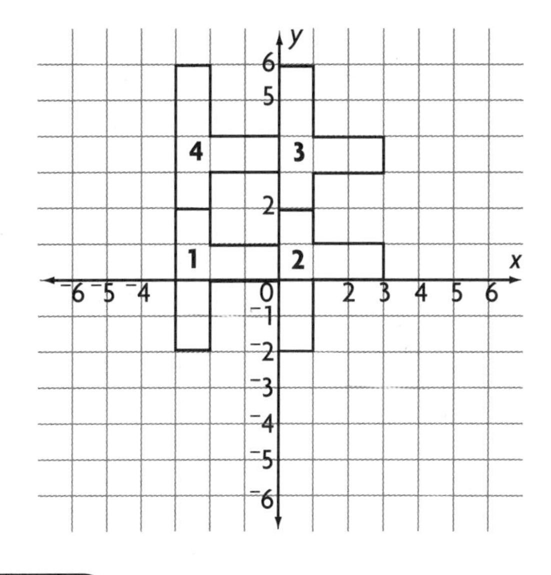

1. In the design at the right, what patterns of transformation are used?

   translation, reflection, translation

2. If you follow the pattern from Problem 1, what transformation will be used to move the figure to position 5?

   reflection

Mixed Applications

Solve.

CHOOSE A STRATEGY

- Write an Equation • Find a Pattern • Make a Chart • Use a Formula • Draw a Diagram • Guess and Check

Choices of strategies will vary.

3. In the design at the right, what transformation would be used to move the figure to position 5? to position 6?

   reflection; rotation

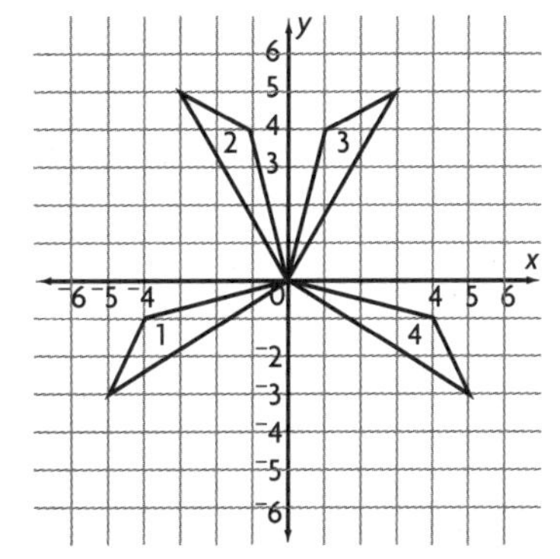

4. Theresa counted the number of mosquitos she saw each day between 5 P.M. and 6 P.M. She saw 1 on the first day, 2 on the second day, 4 on the third day, and 8 on the fourth day. If the pattern continues, how many mosquitos will she count on the fifth day?

   16 mosquitos

5. The perimeter, P, of a certain trapezoid is given by the formula $P = a + b + 2c$. Find the perimeter if $a = 7$ cm, $b = 12$ cm, and $c = 8$ cm.

   35 cm

6. Some two-digit numbers have a units digit that is half the tens digit. Name all the numbers that fit this description.

   21, 42, 63, 84

7. Bill is 3 times as old as Jan. Jan is 2 years older than Charlotte. Charlotte is 5. How old is Bill?

   21 years old

Use with text pages 504–505.

Name ______________________

Patterns of Geometric Figures

Draw the next three figures in each geometric pattern.

1. 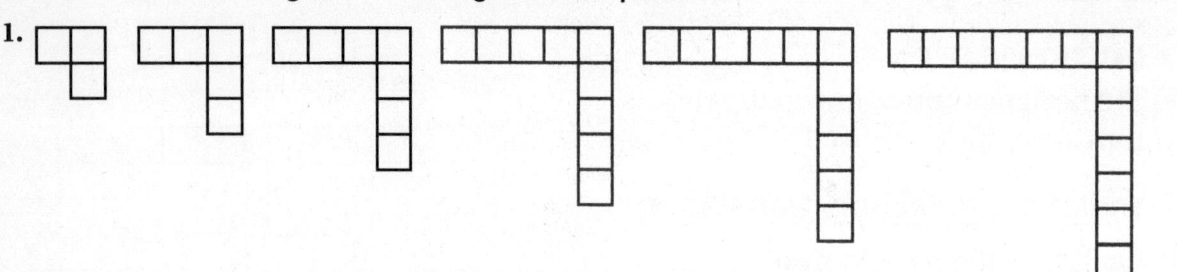

2. 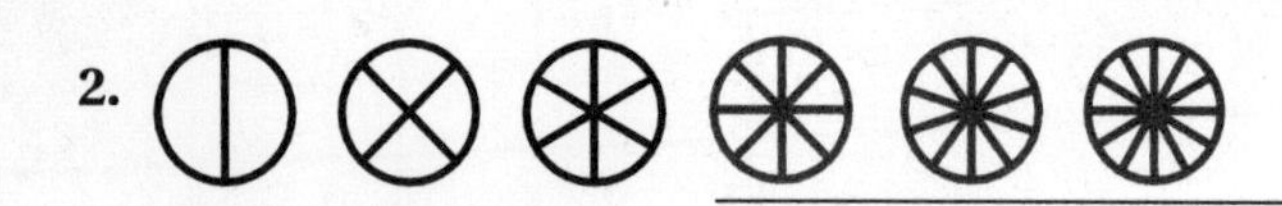

3. 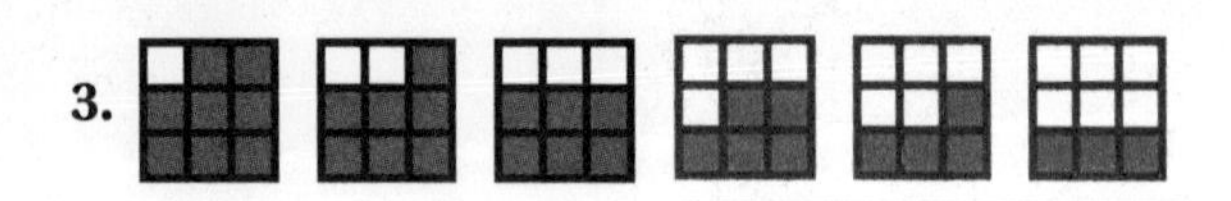

4. 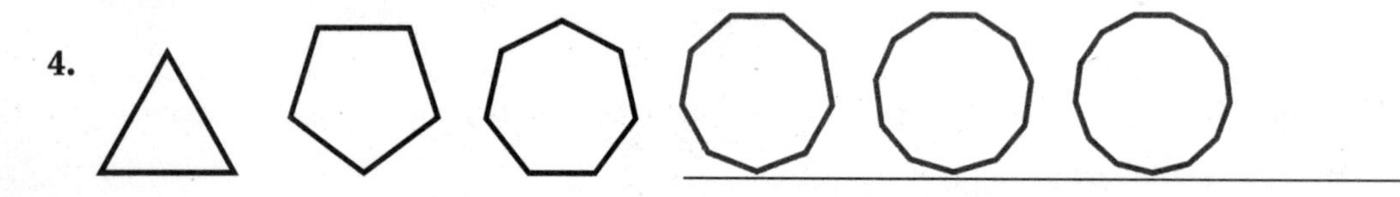

Describe the next two solid figures you would draw for the pattern.

5. Add the same number of cubes to each row and add 1 cube on top.

6. 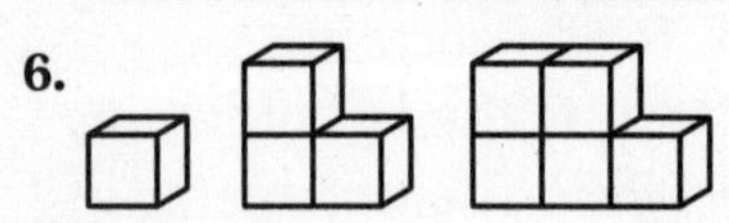

Add 1 cube to each row.

Mixed Applications

7. Boxes of tea are stacked on the shelves of a grocery store. The bottom row has 20 boxes. Each row has one box less than the row below it. There are seven rows. How many boxes of tea are on display?

   119 boxes

8. Jack mows Mr. Hawkins's lawn. The lot is 20 m × 30 m. The house on the lot is 8 m × 10 m. Jack mows the whole lot, except for the part where the house is. How many square meters does he mow?

   520 m^2

Use with text pages 506–507.

Name ______________________________

Making Figures for Tessellations

Make the tessellation shape described by each pattern. Then form two rows of a tessellation. **Check students' tessellations.**

1. 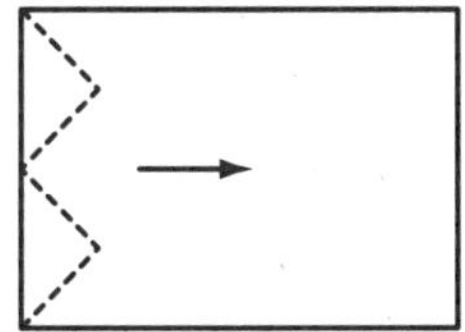

2. 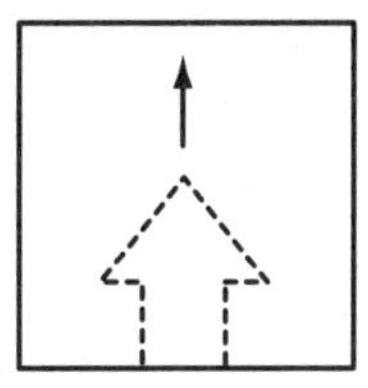

3. 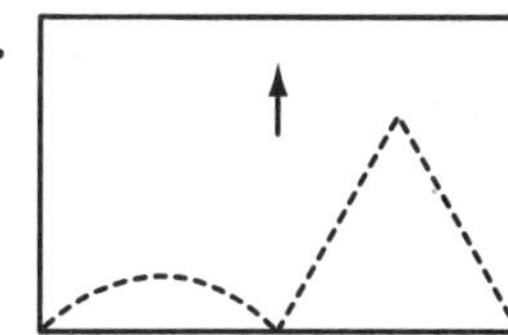

4. 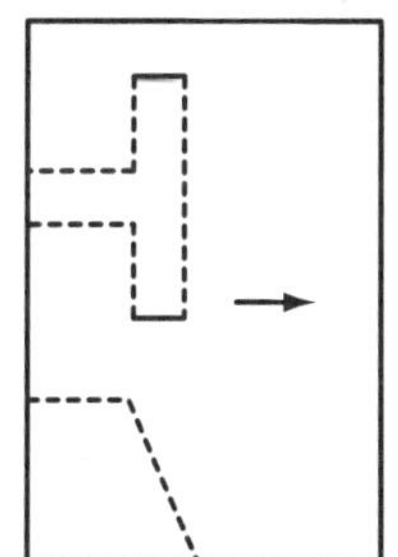

5. 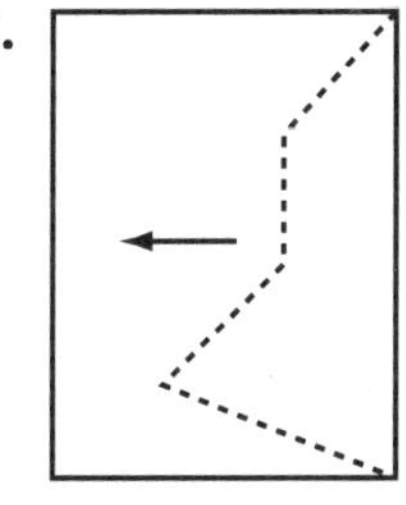

6. 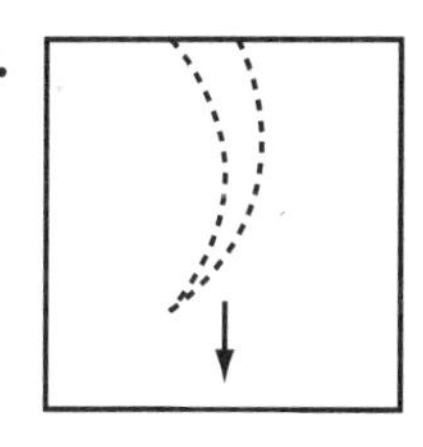

Write *yes* or *no* to tell whether the figure forms a tessellation.

7. 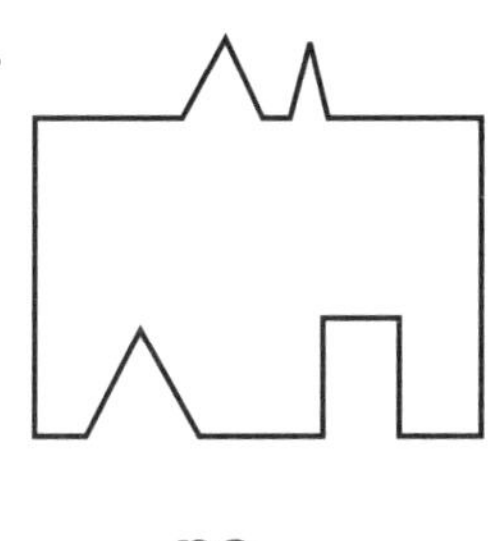

no

8. 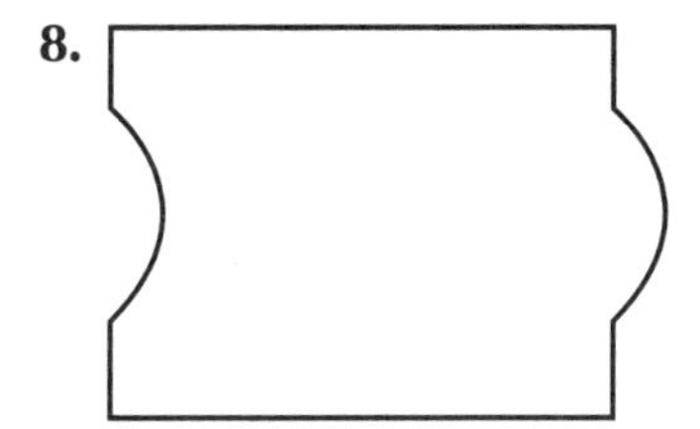

yes

9. 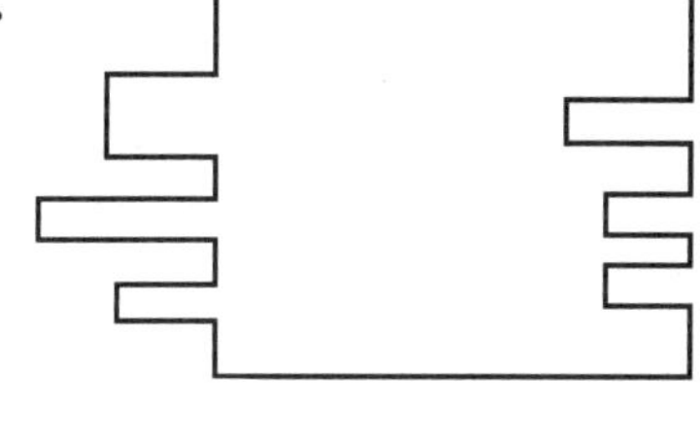

no

Mixed Applications

10. Draw a rectangle. Then mark a part of one side to be cut out. Translate that part to the other side. Then draw at least two rows of the tessellation your design would produce.

Check students' work.

11. A "2-by-4" piece of lumber is really $3\frac{1}{2}$ in. × $1\frac{1}{2}$ in. Find the area of a cross section of a 2-by-4.

$5\frac{1}{4}$ in.^{2}

12. A baker has 5 kg of sugar. She uses 1.8 kg in one recipe and 2.2 kg in another. Does she have enough sugar left for a cake that needs 0.8 kg? Explain.

yes; $5 - 4 = 1$, $1 > 0.8$

13. Four used trucks have the following prices: $15,595; $16,300; $19,500; and $21,650. What are the mean and the median?

mean: $18,261.25;

median: $17,900

Name ______

LESSON 28.1

Number Patterns

Match each term in Column A with its definition in Column B.

	Column A		Column B
1. b	sequence	a.	number of circles used to make a triangular array
2. c	term	b.	an ordered set of numbers
3. a	triangular number	c.	each number in an ordered set of numbers

---

Find the next three terms in the sequence.

4. 17, 22, 28, 35, . . .
   43, 52, 62

5. 81, 69, 57, 45, . . .
   33, 21, 9

6. 1, 5, 25, 125, . . .
   625; 3,125; 15,625

7. 117, 116, 113, 108, . . .
   101, 92, 81

8. 700, 70, 7, 0.7, . . .
   0.07, 0.007, 0.0007

9. 1,000, 500, 250, 125, . . .
   62.5, 31.25, 15.625

10. 77, 79, 83, 85, 89, . . .
    91, 95, 97

11. 19, 16.5, 14, 11.5, . . .
    9, 6.5, 4

12. 64, 55, 47, 40, . . .
    34, 29, 25

13. 17, 34, 68, 136, . . .
    272; 544; 1,088

14. 325, 320, 310, 295, . . .
    275, 250, 220

15. 14.6, 14.5, 14.3, 14.0, . . .
    13.6, 13.1, 12.5

Mixed Applications

16. The Sweet Shoppe made a profit of $150 the first month, $175 the second month, and $200 the third month. If this pattern continues, what will the profit be in the eighth month?

    $325

17. The baseball league has saved $1,000. The league will spend 40 percent of it on new umpire uniforms. Each uniform costs $50. How many uniforms will the league buy?

    8 uniforms

18. Steve has $1.17 in coins. He has a total of 18 coins. What are they?

    11 dimes and 7 pennies

19. Sue saved $40 in January, $55 in February, and $70 in March. If this pattern continues, in what month will Sue save $175?

    October

Use with text pages 516–519.

Name ______________________________

LESSON 28.2

Patterns with Fractions

Find the next three terms in the sequence.

1. $\frac{3}{10}, \frac{7}{10}, 1\frac{1}{10}, 1\frac{1}{2}, \ldots$

   $1\frac{9}{10}, 2\frac{3}{10}, 2\frac{7}{10}$

2. $\frac{5}{6}, 1\frac{1}{6}, 1\frac{1}{2}, 1\frac{5}{6}, \ldots$

   $2\frac{1}{6}, 2\frac{1}{2}, 2\frac{5}{6}$

3. $4\frac{1}{8}, 4\frac{7}{8}, 5\frac{5}{8}, 6\frac{3}{8}, \ldots$

   $7\frac{1}{8}, 7\frac{7}{8}, 8\frac{5}{8}$

4. $8\frac{1}{9}, 7\frac{4}{9}, 6\frac{7}{9}, 6\frac{1}{9}, \ldots$

   $5\frac{4}{9}, 4\frac{7}{9}, 4\frac{1}{9}$

5. $\frac{9}{20}, \frac{13}{20}, \frac{17}{20}, 1\frac{1}{20}, \ldots$

   $1\frac{1}{4}, 1\frac{9}{20}, 1\frac{13}{20}$

6. $8\frac{3}{7}, 6\frac{13}{14}, 5\frac{3}{7}, \ldots$

   $3\frac{13}{14}, 2\frac{3}{7}, \frac{13}{14}$

7. $12\frac{3}{4}, 11\frac{19}{20}, 11\frac{3}{20}, \ldots$

   $10\frac{7}{20}, 9\frac{11}{20}, 8\frac{3}{4}$

8. $2\frac{5}{8}, 2\frac{13}{16}, 3, \ldots$

   $3\frac{3}{16}, 3\frac{3}{8}, 3\frac{9}{16}$

9. $14\frac{1}{2}, 12\frac{1}{4}, 10, \ldots$

   $7\frac{3}{4}, 5\frac{1}{2}, 3\frac{1}{4}$

10. $8\frac{1}{10}, 6\frac{9}{10}, 5\frac{7}{10}, \ldots$

    $4\frac{1}{2}, 3\frac{3}{10}, 2\frac{1}{10}$

11. $5\frac{3}{8}, 5\frac{3}{4}, 6\frac{1}{8}, 6\frac{1}{2}, \ldots$

    $6\frac{7}{8}, 7\frac{1}{4}, 7\frac{5}{8}$

12. $7\frac{2}{3}, 6\frac{5}{12}, 5\frac{1}{6}, \ldots$

    $3\frac{11}{12}, 2\frac{2}{3}, 1\frac{5}{12}$

13. $8\frac{5}{8}, 7\frac{3}{8}, 6\frac{1}{8}, \ldots$

    $4\frac{7}{8}, 3\frac{5}{8}, 2\frac{3}{8}$

14. $12\frac{2}{5}, 13\frac{7}{10}, 15, \ldots$

    $16\frac{3}{10}, 17\frac{3}{5}, 18\frac{9}{10}$

15. $16\frac{1}{8}, 13\frac{3}{8}, 10\frac{5}{8}, \ldots$

    $7\frac{7}{8}, 5\frac{1}{8}, 2\frac{3}{8}$

Mixed Applications

For Problems 16–17, write a sequence, and then solve.

16. Tim spent $\frac{1}{3}$ hr reading on Monday, $\frac{2}{3}$ hr reading on Tuesday, and 1 hr reading on Wednesday. If he continues reading in this pattern, how long will he read on Sunday?

    $\frac{1}{3}, \frac{2}{3}, 1, \ldots; \frac{7}{3}$, or $2\frac{1}{3}$ hr

17. Tina orders plants by their height. The first plant is $17\frac{1}{4}$ in. high, the second plant is $16\frac{1}{8}$ in. high, and the third plant is 15 in. high. If this pattern continues, what is the height of the fifth plant?

    $17\frac{1}{4}, 16\frac{1}{8}, 15, \ldots; 12\frac{3}{4}$ in.

18. Out of every 75 students, 30 play soccer. How many students in a school of 450 play soccer?

    180 students

19. A truck can carry 40 cu ft of soil in one load. How many loads are needed to remove the dirt from a hole that measures 10 ft × 8 ft × 4 ft?

    8 loads

Name ____________________

LESSON 28.3

Patterns with Fraction Multiplication

Find the next three terms in the sequence.

1. $\frac{3}{5}, \frac{6}{5}, \frac{12}{5}, \ldots$

   $\frac{24}{5}, \frac{48}{5}, \frac{96}{5}$

2. $\frac{1}{5}, \frac{1}{10}, \frac{1}{20}, \ldots$

   $\frac{1}{40}, \frac{1}{80}, \frac{1}{160}$

3. $\frac{9}{4}, \frac{9}{12}, \frac{9}{36}, \ldots$

   $\frac{9}{108}, \frac{9}{324}, \frac{9}{972}$

4. $\frac{1}{2}, \frac{1}{8}, \frac{1}{32}, \ldots$

   $\frac{1}{128}, \frac{1}{512}, \frac{1}{2,048}$

5. $\frac{3}{7}, \frac{12}{7}, \frac{48}{7}, \ldots$

   $\frac{192}{7}, \frac{768}{7}, \frac{3,072}{7}$

6. $\frac{3}{4}, \frac{3}{8}, \frac{3}{16}, \ldots$

   $\frac{3}{32}, \frac{3}{64}, \frac{3}{128}$

7. $\frac{3}{8}, \frac{3}{24}, \frac{3}{72}, \ldots$

   $\frac{3}{216}, \frac{3}{648}, \frac{3}{1,944}$

8. $\frac{3}{10}, \frac{3}{5}, 1\frac{1}{5}, \ldots$

   $2\frac{2}{5}, 4\frac{4}{5}, 9\frac{3}{5}$

Write the first four terms of the sequence.

9. pattern: multiply by 6
   first term: $\frac{1}{3}$

   $\frac{1}{3}, \frac{6}{3}, \frac{36}{3}, \frac{216}{3}$

10. pattern: multiply by $\frac{2}{5}$
    first term: $\frac{1}{4}$

    $\frac{1}{4}, \frac{2}{20}, \frac{4}{100}, \frac{8}{500}$

11. pattern: multiply by $\frac{3}{8}$
    first term: $\frac{1}{2}$

    $\frac{1}{2}, \frac{3}{16}, \frac{9}{128}, \frac{27}{1,024}$

12. pattern: multiply by 7
    first term: $\frac{2}{9}$

    $\frac{2}{9}, \frac{14}{9}, \frac{98}{9}, \frac{686}{9}$

13. pattern: multiply by $\frac{1}{9}$
    first term: $\frac{1}{4}$

    $\frac{1}{4}, \frac{1}{36}, \frac{1}{324}, \frac{1}{2,916}$

14. pattern: multiply by $\frac{4}{5}$
    first term: $\frac{1}{10}$

    $\frac{1}{10}, \frac{4}{50}, \frac{16}{250}, \frac{64}{1,250}$

Mixed Applications

15. Kyle is cutting a piece of rope. His first cut resulted in pieces that were $\frac{1}{2}$ the original length. The second cut made pieces that were $\frac{1}{4}$ the original length. The third cut made pieces that were $\frac{1}{8}$ the original length. What part of the original length will the fifth cut yield?

    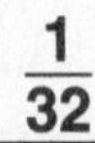

    $\frac{1}{32}$

16. Sue is multiplying a sequence of fractions by 4. Her first term is $\frac{1}{3}$. What is the fourth term in her sequence? Write the term in simplest form.

    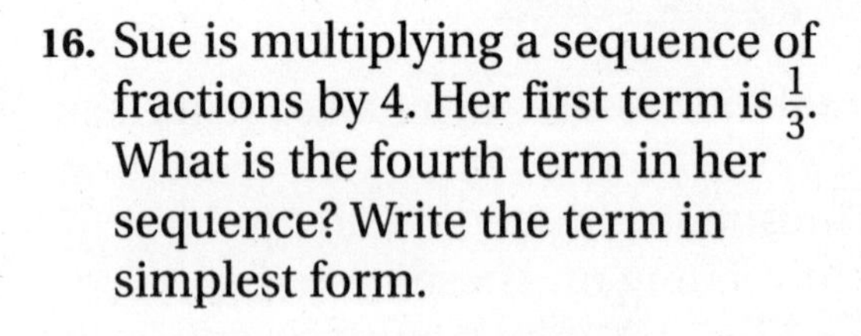

    $21\frac{1}{3}$

17. Out of every 50 students, 15 play a team sport. How many students out of 350 play a team sport?

    105 students

18. The school cafeteria sold 55 lunches on Monday, 68 lunches on Tuesday, and 81 lunches on Wednesday. If this pattern continues, how many lunches will be sold on Friday?

    107 lunches

Use with text pages 524–525.

Name ____________________

Patterns with Integers

Find the next three terms in the sequence.

1. $^{-}24, {}^{-}18, {}^{-}12, \ldots$

   $^{-}6, 0, {}^{+}6$

2. $^{+}5, {}^{+}3, {}^{+}1, \ldots$

   $^{-}1, {}^{-}3, {}^{-}5$

3. $^{+}400, {}^{+}200, 0, \ldots$

   $^{-}200, {}^{-}400, {}^{-}600$

4. $^{-}32, {}^{-}24, {}^{-}16, \ldots$

   $^{-}8, 0, {}^{+}8$

5. $^{+}7, {}^{+}2, {}^{-}3, \ldots$

   $^{-}8, {}^{-}13, {}^{-}18$

6. $^{-}11, {}^{-}8, {}^{-}5, \ldots$

   $^{-}2, {}^{+}1, {}^{+}4$

7. $^{+}4, {}^{-}8, {}^{+}16, \ldots$

   $^{-}32, {}^{+}64, {}^{-}128$

8. $^{-}3, {}^{+}9, {}^{-}27, \ldots$

   $^{+}81, {}^{-}243, {}^{+}729$

9. $^{+}4, {}^{-}20, {}^{+}100, \ldots$

   $^{-}500, {}^{+}2{,}500, {}^{-}12{,}500$

10. $^{-}700, {}^{+}70, {}^{-}7, \ldots$

    $^{+}0.7, {}^{-}0.07, {}^{+}0.007$

11. $^{-}13, {}^{+}26, {}^{-}52, \ldots$

    $^{+}104, {}^{-}208, {}^{+}416$

12. $^{-}0.004, {}^{+}0.04, {}^{-}0.4, \ldots$

    $^{+}4, {}^{-}40, {}^{+}400$

13. $^{+}4, {}^{-}2, {}^{+}1, \ldots$

    $^{-}0.5, {}^{+}0.25, {}^{-}0.125$

14. $^{-}8, {}^{+}40, {}^{-}200, \ldots$

    $^{+}1{,}000, {}^{-}5{,}000, {}^{+}25{,}000$

15. $^{-}250, {}^{+}50, {}^{-}10, \ldots$

    $^{+}2, {}^{-}0.4, {}^{+}0.08$

16. $^{-}100, {}^{+}20, {}^{-}4, \ldots$

    $^{+}0.8, {}^{-}0.16, {}^{+}0.032$

17. $^{+}12, {}^{-}36, {}^{+}108, \ldots$

    $^{-}324, {}^{+}972, {}^{-}2{,}916$

18. $^{-}1{,}000, {}^{-}100, {}^{-}10, \ldots$

    $^{-}1, {}^{-}0.1, {}^{-}0.01$

19. $^{-}23, {}^{-}18, {}^{-}13, \ldots$

    $^{-}8, {}^{-}3, {}^{+}2$

20. $^{+}0.5, {}^{-}2.5, {}^{+}12.5, \ldots$

    $^{-}62.5, {}^{+}312.5, {}^{-}1{,}562.5$

21. $^{+}12{,}500, {}^{-}500, {}^{+}20, \ldots$

    $^{-}0.8, {}^{+}0.032, {}^{-}0.00128$

22. $^{-}17, {}^{-}14, {}^{-}11, \ldots$

    $^{-}8, {}^{-}5, {}^{-}2$

23. $^{-}1.5, {}^{+}3, {}^{-}6, \ldots$

    $^{+}12, {}^{-}24, {}^{+}48$

24. $^{+}300, {}^{-}30, {}^{+}3, \ldots$

    $^{-}0.3, {}^{+}0.03, {}^{-}0.003$

Mixed Applications

25. Today the temperature is 76°. The forecast calls for a decrease of 3° for each of the next four days. What is the predicted temperature four days from now?

    64°

26. The first water sample is taken from the ocean at $^{-}60$ m. Other samples are taken at intervals of 40 m below the first one. From what depth is the fifth sample taken?

    $^{-}220$ m

27. A service call to repair a refrigerator costs \$40 plus \$25 per hour. Scott's refrigerator took $1\frac{1}{4}$ hr to fix. How much did the repair cost?

    \$71.25

28. Of a group of 160 customers, 40 percent charged their purchases. How many members of the group did not charge their purchases?

    96 customers